Smart Innovation of Web of Things

Internet of Everything (IoE)

Series Editors

Vijender Kumar Solanki,
Raghvendra Kumar and Le Hoang Son

IoT: Security and Privacy Paradigm
Edited by Souvik Pal, Vicente Garcia Diaz and Dac-Nhuong Le

Smart Innovation of Web of Things
Edited by Aarti Jain, Rubén González Crespo and Manju Khari

Smart Innovation of Web of Things

Edited by
Aarti Jain, Rubén González Crespo and
Manju Khari

CRC Press
Taylor & Francis Group
Boca Raton London New York

CRC Press is an imprint of the
Taylor & Francis Group, an **informa** business

First edition published 2020
by CRC Press
6000 Broken Sound Parkway NW, Suite 300, Boca Raton, FL 33487-2742

and by CRC Press
2 Park Square, Milton Park, Abingdon, Oxon, OX14 4RN

ISBN: 978-0-367-27565-5 (hbk)
ISBN: 978-0-429-29846-2 (ebk)

Typeset in Palatino
by Deanta Global Publishing Services, Chennai, India

Contents

Preface

The focus of this book is the area of the Web of Things (WoT), which has witnessed revolutionary advancements in the last few months. Web of Things defines different approaches, programming tools and architectural styles that connect computationally effective real-world objects, i.e. IoT with the World Wide Web. In other words, much as an application layer is to the network layer of the internet, the Web of Things is an application layer that helps in the simplification of the creation of applications of the Internet of Things (IoT). The Web of Things does not relate to new standards, but rather re-uses existing well-known web standards, viz. the semantic web (e.g. microdata, JSON-LD, etc.), the social web (e.g. social networks or OAuth), the real-time web (e.g. WebSockets) and the programmable web (e.g. JSON, REST, HTTP). The concept of the Web of Things itself is evolving and currently a lot of work is being done in this field. The developments in this sector impact our lives in our homes, workplace and our way of working, studying, transacting and entertaining.

In the near future, WoT is likely to be applied in a wide range of fields, such as environmental monitoring, healthcare, transportation, smart cities, smart homes, urban planning, infrastructure monitoring and agriculture. Collective development in the fields of ubiquitous computing, social networks, cyber security, digital, social and physical framework circle assembly is also likely to be a major area of research. This book is organized into 11 chapters, each focusing on a unique facet of wireless technological aspects of the Web of Things, and it aims to comprehensively cover its various applications.

Chapter 1 notes the promising opportunity that has been provided by Web of Things (WoT) for building powerful applications and systems using radio-frequency identification (RFID), mobile, wireless and sensor devices. It covers a large number of applications based on WoT architectures that have been developed in recent years. The main purpose of this chapter is to understand how WoT technology works. It also reviews the key technologies, WoT protocols and WoT applications in different areas. In the end, it summarizes the complete WoT system.

Chapter 2 offers an insight into the history of the Web of Things and delay of networks. This chapter focuses on how the Web of Things is the actual implementation of the internet web concept (the World Wide Web). It highlights a myriad of applications of delay-tolerant networks (DTNs) that experience the mobility issues of lack of continuous links between the nodes and transient topology. This includes, but is not limited to, wildlife tracking networks, monitoring scientific and hazard events, and providing comms systems and data transfer facilities to remote rural sites, underwater networks and interplanetary networking (IPN). The chapter also addresses a multitude of DTN routing strategies that tackle the provision of transporting and steering bundles from source node to target.

Chapter 3 sheds light on the advancements in modern wireless communications that enhance internet access to a connected network (IoT), which in turn explores the data from IoT with a connectivity service and applications through Web of Things (WoT). It also discusses the major role WoT occupies in smart city development. It focuses on the factors that make WoT challenging and how the researchers are attracted toward handling these challenges. Apart from various smart infrastructure, the chapter explains how smart energy management is a good alternative to developing a smart city. The chapter also proposes the integration of WoT in smart buildings for energy management using a deep learning dashboard for decision-making from sensor data.

The concern of Chapter 4 is epileptic seizures which are sudden changes in human behavior due to abnormal electrical activity in the human brain, which leads to uncontrollable human bodily activity. This chapter emphasizes various state-of-the-art techniques that have been proposed by various research groups based on electroencephalogram (EEG) signals' feature extraction, followed by classification. This chapter is mainly based on the design and development of a non-invasive method to predict, classify and detect epileptic seizures. Moreover, it uses wireless technology that will be paired with external control devices. The outcome of this chapter will provide a real-time alert system for monitoring epileptic seizure prediction and detection, which will be beneficial to society.

Chapter 5 focuses on the frequency spectrum, which is considered to be the most pivotal, yet limited, natural resources. It also introduces the rapid utilization of wireless technology in communications. The chapter presents advanced applications, including multimedia communication, telemedicine, smart spaces, smart cities and many more. In addition, it explains how cognitive radio technology is better than the present mobile communication technology. It is also based on the benefits of cognitive radio networks (CRNs).

Chapter 6 emphasizes notable factual attempts that have been utilized in the build-out of statistical methods which could provide potent and efficient botnet detection, prevention and mitigation. It highlights network traffic, which is the key principle for discovering botnet existence and many ultramodern pathways that use machine learning methodologies and techniques for discovering malicious traffic. The chapter presents a brief study of botnets' life cycles and modern detection methods for identifying botnet network traffic.

Chapter 7 covers an intelligent solution to tackle critical issues, such as healthcare, energy management, transportation and infrastructure. It discusses the importance of the Internet of Things in tackling these challenges through automation, networking and sensing and proper data analysis. It explains new technologies and connected data sensors using wireless or wired communication, which are utilized by smart cities. The chapter is devoted to technologies like artificial intelligence, R programming, Python and machine language which is capable of helping the network in processing the information received from the connected gadgets. Moreover, it presents the major use case of a smart city being smart transportation systems, smart parking, smart building monitoring, smart agriculture, smart waste management and smart security systems.

The objective of Chapter 8 is to present the involvement of the Web of Things in retail management. It offers a discussion of new technologies related to the retail industry, which provides better services to customers and retailers. It introduces some of the benefits enjoyed by retailers as a result of WoT. There is also a focus on proficient technologies provided by WoT to access, store, share and investigate a massive amount of data which is quickly generated on a daily basis. Contributions of WoT in supply chain and logistics in retail management are discussed here as well. The chapter is based on the security issues that the WoT-enabled retail industry is currently facing. The chapter presents the concept of the Web of Things as it relates to infrastructure, its applications and challenges in retail management.

Chapter 9 describes the necessity of the banking sector for the socio-economic growth of any country. It introduces the challenges associated with the banking sector by the GAFAs (Google, Apple, Facebook and Amazon) and by the companies offering/developing various pieces of financial software. It offers a brief introduction to the fundamental task of any banking sector in an economy. The chapter sheds light on the principles of WoT to automate some tasks which were initially tedious for humans to do efficiently. It describes the potential of WoT to completely modify the way the banking sectors work. This chapter is organized as follows: benefits of WoT in the banking sector, examples of WoT in banking and financial services, challenges and design issues with WoT-enabled banking services, and the growth rate of WoT-enabled banking services.

The main aim of Chapter 10 is to focus on some of the difficulties retailers face from various points, for example, disintegrating product margins, elevated challenges and the consistent strain to improve. It discusses the advancements of the Web of Things to accomplish a genuine omnichannel experience by carrying digital innovations to physical stores. It also concludes that customers should be given opportunities for interfacing with retailers crosswise over different channels for finishing their shopping ventures, which may include a choice of on the web and offline blends. The chapter looks at how WoT could affect retail, the opportunities and difficulties ahead, and what one has to do to get a WoT system all together.

Chapter 11 is concerned with the alarming growth rates of malicious applications that pose a grave issue. It presents a study which analyzes that in every 10 seconds a new malware application is introduced in Android. It includes a discussion on requirements of a scalable malware detection approach, which can efficiently figure out malicious apps from a pool of applications of both harmful and benign apps. It also emphasizes innumerable tools for malware identification that have been introduced at the system level as well as the network layer. The authors propose an automated tool that would extract features from Application Package Kit (APK) files and form a dataset of it, which could then be used for static analysis of the data. Moreover, it focuses mainly on the proposed automated tool that would create a dataset by reading the APK files and extracting features mainly from two files, i.e. Manifest.XML and Classes.dex. It also describes how this tool can be helpful for researchers to create their own dataset.

Aarti Jain
Rubén González Crespo
Manju Khari

MATLAB® is a registered trademark of The MathWorks, Inc. For product information, please contact:

The MathWorks, Inc.
3 Apple Hill Drive
Natick, MA 01760-2098 USA
Tel: 508 647 7000
Fax: 508-647-7001
E-mail: info@mathworks.com
Web: www.mathworks.com

Editors

Aarti Jain, PhD, is an Assistant Professor in AIACTR, Under Government of NCT Delhi, affiliated with GGSIP University, Delhi, India. She holds a PhD in communications technologies from GGSIP University and received a master's degree in electronics and communications from Delhi College of Engineering, University of Delhi, in 2009. Her research interests include the Internet of Things, energy and quality management for wireless sensor networks, bio-inspired computing and its application and localization technologies.

Rubén González Crespo, PhD, holds a PhD in computer science engineering. Currently, he is Dean of the School of Engineering and Technology from UNIR and Director of Policy and Academic Planning. He is Editor in Chief of the *International Journal of Interactive Multimedia and Artificial Intelligence,* and editor for several indexed journals. His main research areas are soft computing, accessibility and TEL. He is an advisory board member for the Ministry of Education in Colombia and an evaluator for the National Agency for Quality Evaluation and Accreditation of Spain (ANECA). He has published more than 180 papers in indexed journals.

Manju Khari, PhD, is an Assistant Professor in AIACTR, Under Government of NCT Delhi affiliated with GGSIP University, Delhi, India. She is also the Professor-in-Charge of the IT Services of the Institute and has more than 12 years of experience in network planning and management. She holds a PhD in computer science and engineering from the National Institute of Technology Patna and received her master's degree in information security from AIACTR (formerly Ambedkar Institute of Technology affiliated with GGSIP University, Delhi). Her research interests are software testing, software quality, software metrics, information security and nature-inspired algorithms.

Contributors

Siddhant Banyal
Department of Instrumentation and
  Control
Netaji Subhas University of
  Technology
New Delhi, India

Kartik Krishna Bhardwaj
Department of Instrumentation and
  Control
Netaji Subhas University of
  Technology
New Delhi, India

Amit Bhati
Department of Information
  Technology IET
Dr. Rammanohar Lohia Avadh
  University, Ayodhya
Uttar Pradesh, India

Kirti Dalal
Department of Electronics and
  Communication Engineering
Ambedkar Institute of Advanced
  Communication Technologies and
  Research
New Delhi, India

Renu Dalal
Computer Science and Engineering
  Department
AIACT&R, GGSIP University
Delhi, India

Tanuja S. Dhope (Shendkar)
JSPM's Rajarshi Shahu College of
  Engineering, Tathawade
Pune, India

Aditya Garg
Department of Computer Science
AIACT&R, GGSIP University
Delhi, India

Ashish Gupta
Department of Computer Science &
  Engineering
Dr. Rammanohar Lohia Avadh
  University, Ayodhya
Uttar Pradesh, India

Aarti Jain
Department of Electronics and
  Communication Engineering
Ambedkar Institute of Advanced
  Communication Technologies and
  Research
New Delhi, India

Vijayalaxmi Jain
Dr. D Y Patil International University
  Akurdi
Pune, India

V. Jeyabalaraja
Department of CSE
Velammal Engineering College
Chennai, India

M. Kaliappan
Department of CSE
Ramco Institute of Technology
Rajapalayam, India

Manju Khari
Department of Computer Science
AIACT&R, GGSIP University
Delhi, India

S. Koteeswaran
Department of Computer Science
and Engineering, School of
Computing
Dr. Sagunthala R&D Institute of
Science and Technology
Chennai, India

Ashish Kumar
Department of Computer Science
AIACT&R, GGSIP University
Delhi, India

Udit Misra
Department of Computer Science
AIACT&R, GGSIP University
Delhi, India

Deepti Mishra
G.L. Bajaj Institute of Technology &
Management
Uttar Pradesh, India

Rupali Rani
Ambedkar Institute of Advance
Communication Technology &
Research
New Delhi, India

Pratibha Rohilla
Computer Science and Engineering
Department
AIACT&R, GGSIP University
Delhi, India

Deepak Kumar Sharma
Department of Information
Technology
Netaji Subhas University of
Technology
New Delhi, India

Bahubali Shiragapur
Dr. D Y Patil School of Engineering,
Lohegaon
Pune, India

Dina Simunic
FER
University of Zagreb
Zagreb, Croatia

P. Subbulakshmi
Department of CSE
Hindustan Institute of Technology
and Science
Chennai, India

A. Suresh
Department of CSE
Nehru Institute of Engineering and
Technology
Coimbatore, India

Nishikant Surwade
Dr. D Y Patil School of Engineering,
Lohegaon
Pune, India

S. Vimal
Department of Information
Technology
National Engineering College
Kovilpatti, India

1

Emergence of the Web of Things: A Survey

Aditya Garg and Manju Khari

CONTENTS

A promising opportunity has been provided by the Web of Things (WoT) for building powerful applications and systems by using radio-frequency identification (RFID), mobile, wireless and sensor devices. A large number of applications based on WoT architectures have been developed in recent years. WoT systems will help in the emergence of a cyber-world from the existing physical world and will ultimately result in changing human interaction with the world. The main purpose of this chapter is to understand how WoT technology works. This chapter reviews the key technologies, WoT protocols and WoT applications in different areas, and summarizes the complete WoT system.

1.1 Introduction

The first version of the Electronic Product Code (EPC) network was launched by the Auto-ID Centre for the identification and detection of the supply of goods in supply chains. This was the first time that WoT came to attention. WoT is treated as a refinement of the IoT (Internet of Things) architecture. IoT was first mentioned in a paper of the Auto-ID Center that was about Electronic Product Code and was written by David Brock in 2001 [1]. Afterward, WoT started to be considered as a future value that was essential for the internet and was important for the realization of machine-to-machine learning [2].

One cannot find a universal definition that can truly define what WoT is; thus a better approach is to define the core concept of WoT. The core concept for WoT defines it in such a way that it is not only the objects that are used in day-to-day life that can be equipped with sensors, networking and processing capabilities that will in turn enable the objects to intercommunicate, but these real world objects can also be included in the architecture as well. Although the technology behind this core concept is not new, as these technologies involve the use of RFID and sensors in the context of industries and manufacturing for the detection of large ticket items. Only the evolution of existing methods such as machine-to-machine learning is represented by WoT. WoT also represents the interconnection of devices, as well as their interconnection over the network.

According to a study by the GSMA, the number of devices that were interconnected over the WoT network has overtaken the actual number of people on the earth. According to current criteria, about 9 billion devices are connected to each other, and this is expected to increase further to 24 billion. Moreover, approximately $1.3 trillion revenue opportunities are generated for mobile network operators.

For the correct functioning of Web of Things, the basic demands that need to be fulfilled are as follows:

1. There must be an understanding between users and the applications.
2. There must be software architecture and pervasive communication networks present in order to processing and convey relevant contextual information.

When these fundamentals are successfully met, context-aware computation and smart connectivity can easily be accomplished.

Nowadays, the user base of the internet is increasing at a rapid speed. Billions of people all around the globe use the internet for performing various activities including surfing, sending and receiving emails, online gaming and many other tasks. With an increasing population, there is also a rapid increase in the number of people who gain access to communication infrastructure and global information. Within such a perspective, WoT refers to three things: First, smart objects are interconnected over the global network with the help of extended technologies. Second, the requirement of supporting technologies is increasing and is also necessary for the realization of the vision. Third, the applications and its services which are using these technologies help in opening up new markets and business opportunities [3].

1.2 Methodology

WoT is becoming more and more common. WoT can be used as a reference for almost anything that is smart, such as smartphones and smart homes. WoT will play a major role in changing the world by making almost everything free from human interaction. In a WoT environment, the applications are designed in such a way that they can imitate humans and work on their behalf. Various applications are being developed by researchers that are automating human tasks.

To fully understand WoT, it is required to completely understand the architecture of WoT. In comparison with other surveys, this chapter presents a deeper summary of the relevant protocols and standards that are used in WoT to enable researchers for speeding things up. The authors provide some key challenges, vulnerabilities and attacks that WoT has faced in recent years. In this chapter, various security and privacy techniques are also described to help protect data in WoT. Moreover, this chapter explores the relationship between WoT and the latest emerging technologies that are included in cloud computing, and so on.

1.3 WoT Architecture

This section describes the architecture that forms the basis of WoT. For an explanation of architecture of WoT, it is not really necessary to delve into the electronic and hardware parts of it. This section will go into further depth on the software stack and how software plays an essential role in our WoT system. WoT structure is basically distributed into five layers, as presented in Figure 1.1. These layers are as follows:

1. *Perception layer*: This can also be termed the recognition layer [4]. This layer is the lowermost and most basic layer that is included in the architecture of WoT. Both physical objects as well as sensor devices are present in this layer. The main responsibility that the layer performs is the collection of useful data from things like other devices and sensors, and also its transformation in a digital setup. Basically, its responsibility is the identification and also the collection of object-specific information. The major purpose is unique address identification and communication among technologies which are short-range, such as Bluetooth, RFID, NFC and Low Power Personal Area Network (6LoWPAN) [5].

2. *Network layer*: The other name for this layer is "transmission layer." As the name suggests this layer is used in the transmission of data between the sensor devices. This layer provides a secure passage for the transfer of sensitive information obtained from sensor devices to the system that processes information. The mode of transmission can be wireless, such as UMTS, Wi-Fi 3G, infrared, etc., and it can also be wired. Thus, the information is transferred from the perception layer to the middleware layer with the help of network layer.

3. *Middleware layer*: Each device connects only to those specific devices that implement common services. The responsibility of this layer is the linkage and management of such devices and also linking them to the database. The network layer sends information here, and it is stored in the database. The various functions performed by this layer are information processing and ubiquitous computing. With the help of these functions, this layer makes automatic decisions.

4. *Application layer*: This layer is the topmost layer of conventional WoT architecture. Depending on the need of the user, this layer can be

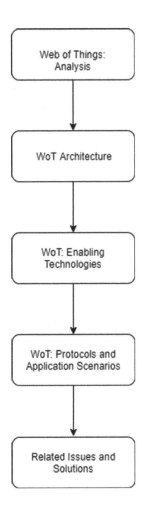

FIGURE 1.1
WoT: Methodology.

modified to provide personalized services [4]. All the Real-World Things reside in this layer of the WoT architecture. Network Things such as Wi-Fi, Bluetooth, Ethernet etc. are present in this layer. This layer fulfils the role of linking users and application to each other. Different types of application-based solutions such as health and disaster monitoring are handled by this layer in combination with the business layer [6].

5. *Business layer*: This layer is responsible for the management of WoT system including applications and their services as well. The success of WoT technology is dependent on a good business model. On the basis of result analysis, this layer helps in the determination of future actions and business strategies [6, 7].

1.3.1 SOA-Based Architecture

Service-oriented architecture (SOA) is defined as a services collection that can intercommunicate with the help of standardized interaction patterns. There are two types of communication that are possible. First, passing of simple messages, and second, coordination of two or more services with the help of appropriate protocols. Presently, usage of many SOC (security operations centers) deployments is through protocols based on the web (e.g. http), so that interoperability is supported across enabling technologies and administrative domains. Web services are managed with the help of services to make them behave like a virtual network, adapting applications depending on the user's need. Only the provided level of heterogeneity and flexibility that are to be deployed in the software modules are supported by service-oriented architecture [8]. Straightforward construction of WoT applications is not possible using SOC/SOA architecture.

One of the most formidable approaches used in service-oriented architecture is that architecture can be created on the basis of system services. Utilization of middleware is a key approach in SOA. The middleware concept can be explained as a software layer that is superimposed between the technology and application layers. With the superimposition of this layer, unnecessary details are hidden and hence the time required for developing the product is reduced [9].

The involvement of RFID with the SOA-based architecture gave rise to many more new possibilities. RFID-SN (RFID-enabled Sensor Network) has been developed by researchers and comprises an RFID tag, a reader and a computer-based system for understanding its behavior [10]. The SOA paradigm has been utilized by scientists for developing a RFID-based system. These systems make use of multiple data-related services such as filtering, aggregation, tag identifier, and so on [11, 12].

1.4 WoT Technologies

1.4.1 Radio-Frequency Identification (RFID)

RFID is defined as a system that wirelessly and with the help of radio waves causes the identity transmission of a person or object as a serial number [13]. In 1948, the first RFID device was created and was used for identification of friends or foes of Britain in the World War II. After that, the evolution of RFID technology started in 1999 at MIT. For tackling issues of identification of objects in a cost-effective manner, RFID technology plays a significant role [14]. There are three types of RFID technologies:

1. Active RFID
2. Passive RFID
3. Semi-Passive RFID

1.4.2 Near-Field Communication (NFC)

NFC is a short-range wireless technology which can only work up to a distance of 4 cm. This technology works in a frequency range of 13.56 MHz. With the help of NFC technology, a touch between two devices can help in making transactions, exchanging digital content and much more. NFC technology was initially developed by the Philips and Sony companies and the advantages of these technologies include working in even dirty environments, no requirement of line of sight and simple connection methods. Moreover, only under 15 ma of power is consumed by NFC [15].

1.4.3 Bluetooth

Bluetooth is another short-range technology. It is inexpensive and it also helps in the elimination of wired connection between devices. The effective range for Bluetooth is 10–100 m. The specification used by the IEEE is 802.15.1 and the communication speed is less than 1 Mbps. The extended functionality of Bluetooth is called Piconet. In Piconet, a common communication channel is shared by a set of Bluetooth devices. Piconet can at maximum connect 2–8 devices at a single time and the can be used to share any kind of data in between them [15].

1.4.4 Wireless Fidelity (Wi-Fi)

Wi-Fi is another networking technology that allows device intercommunication. Initially when wireless products came into the market, they only supported speeds of 1-1 Mbps and were sold under the name of WaveLAN. Today, Wi-Fi devices deliver high speed over WLAN. Nowadays, Wi-Fi is used almost every place including offices, homes and public locations. Wi-Fi technology includes different types of WLAN products and it supports IEEE 802.11 and can also support dual-band including 802.11a, 802.11b, 802.11g and 802.11n.

1.4.5 Long-Term Evolution Advanced (LTE-A)

LTE-A is defined and treated as a set of cellular communication protocols that are used for Machine-Type Communications (MTC). This type of communication is also used in smart cities for achieving long-term durability [16].

Orthogonal Frequency Division Multiple Access (OFDMA) is used in the physical layer by LTE-A so that the channel bandwidth can be easily partitioned into much smaller bands that are typically known as physical resource blocks (PRBs). Multi-component carrier is also implemented by LTE-A which allows it to have up to five 20-MHz bands. There are two parts to the LTE-A architecture:

1. *Core Network*: It controls the mobile devices and also takes care of the IP packet flows.
2. *Radio Access Network (RAN)*: The responsibility of this part is handling the wireless communication as per the user and control plane. It consists of many base stations that are well-connected with the help of the X2 interface.

1.4.6 Wireless Sensor Networks (WSNs)

WSN uses multi-layered protocols and it consists of finite sensor nodes which are called motes. The sensor nodes are controlled by a single special purpose node which is called a sink [17,18]. IEEE 802.15.4 is incorporated by most WSN systems for WPAN communication purposes. Ipv6 addressing functionality is provided by protocol stacks which enhance the ability to control a large number of nodes [19]. The e-SENSE project [20] makes use of WSN with the help of a layered logical approach for providing intelligent user support [21]. UbiSec&Sens [22] is another example of a system that make use of WSN and is very much similar to e-SENSE. The major difference between the two is that UbiSec&Sens has another security layer added on the top.

1.4.7 Cloud Computing

Cloud computing provides the client systems with the platform, infrastructure and software as a service for managing data, along with accessing and processing, which is expressed in the form of pay-as-you-go service or a free service [23, 24]. The cloud infrastructure also provides vehicle-based cloud data services which are used for incorporating intelligent parking cloud services [25].

1.4.8 Zigbee

ZigBee is another wireless technology that is used in WoT which operates on 2.4 GHz spectrum. The data rate of this type of wireless technology is limited to only 250 kbps and has a range of up to 100 m. ZigBee is defined as a mesh network protocol and not all the devices can be used as bursts; moreover, depending on the device position, it determines whether they are used as router or as a controller within the mesh.

1.4.9 Big Data Analytics

WoT architecture relies on a large volume of data. This large volume of data is increasing at a rapid rate. Hence, there is a need for managing such data. Big Data Analytics proposes data storage techniques for storage of both structured and unstructured data in a WoT environment. The architecture makes use of "Hadoop" and many other databases for the creation of a distributed file repository for storing and managing various types of data which are collected by RFID readers and sensors [26]. Another problem arises in managing the large data volume which is generated by sensors and nodes that are present in the WoT-based systems [27].

1.5 WoT Protocols

1.5.1 Internet Protocols

Internet protocols (IP) were developed in 1970s and are defined as the principal network protocol. To relay datagrams across network boundaries, the most important communication protocol is the internet protocol. IPv4 and IPv6 are different types of IPs that are used in computer networks. Ipv4 is a 32-bit address that is used for unique identification of objects and computers. IPv6 is generally a 128-bit address. There are about five different classes that are present for IP ranges inIPv4, namely, Classes A–D, and lastly Class E. Among all of these classes only Class A, B and C are commonly used. According to the actual protocols there are around 4.3 billion IPv4 addresses while the availability of IPv6 is significantly more – around 85,000 trillion [28]. IPv6 supports around 2,128 addresses [15].

1.5.2 Application Protocols

This section discusses different types of application protocols. Figure 1.2 describes various application protocols that are present in WoT (Figure 1.3).

1.5.3 Constrained Application Protocol (CoAP)

The CoAP operates on the application layer and was created by IETF Constrained RESTful Environments (CoRE) [29]. CoAP also consists of web transfer protocols which are defined on the basis of HTTP and REST. REST defines a simpler way for data transfer between the client and the server [30]. For meeting WoT requirements, the HTTP functionalities were modified by using CoAP.

The main purpose of CoAP is enabling RESTful interactions in low power devices. There are two sublayers into which CoAP can be divided as follows: First, the messaging sublayer that helps in the detection of duplicates and also

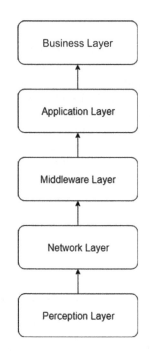

FIGURE 1.2
WoT architecture – Five layers.

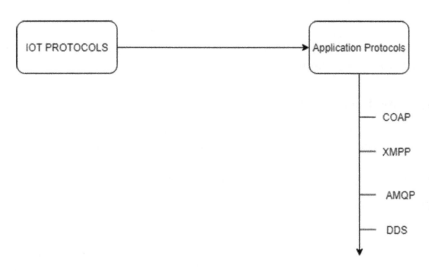

FIGURE 1.3
WoT protocols.

helps in providing reliable communication over the UDP transport layer. The second is the request/response sublayer that handles REST communications [31].

1.5.4 Extensible Messaging and Presence Protocol (XMPP)

XMPP is defined as an IETF instant messaging (IM) standard that is used for multi-person chatting, voice calling and also for video calling [32]. Because of the features that are present in XMPP, it makes the protocol preferable for most of the IM applications, and it is also relevant when concerned within the scope of WoT.

On the top of the core products, XMPP is secure and it also allows new products to be added. There is a high network overload due to the text-based communication in XMPP. A simple answer to this kind of problem is to compress XML with the help of EXI [31].

1.5.5 Advanced Message Queueing Protocol (AMQP)

AMQP [32] is an open source protocol with a main focus on environments that are message based. Trusted communication is supported by AMQP due to the presence of message delivery guarantee primitives. A trusted transport protocol, such as TCP, is required by AMQP to exchange messages.

With the help of wire level protocol, implementations based on AMQP are able to interoperate with each other. AMQP not only supports point-to-point communications, but also supports communication models such as publish/subscribe [31].

1.5.6 Data Distributed Service (DDS)

Data Distributed Service was developed by Object Management Group and is based on publish-subscribe models. These models are used for real-time mobile-2-mobile communications [33]. In comparison with other protocols such as AMQP or MQTT, an excellent service quality is presented by DDS with the help of broker-less architecture. Twenty-three QoS policies are supported by DDS. By using these 23 QoS policies, different communication criteria such as urgency, security, reliability and durability can be addressed by developers.

1.6 WoT Applications

1.6.1 Home Automation

Home automation plays a very important role in many WoT applications. With the help of home automation, smart cities can be realized. WoT has played an important role in the establishment of smart homes. Some of the important features like remote monitoring and controlling of electrical applications can be performed easily by smart homes. The development of smart cities as proposed by the present Indian government [34] all over the country will result in a huge demand for smart home creation.

1.6.2 Automotive

With the help of advanced sensors and actuators, the advancement of vehicles such as buses, trains, and so on, as well as bicycles is booming. With the help of smart things, applications that are present in the automotive industry can now be used for monitoring and reporting various parameters that can go from a small thing such as tire pressure to big things like vehicle proximity. RFID has been equipped to streamline vehicle production, increase quality and improve customer service. Intelligent Transportation Systems (ITS) such as vehicular safety and management of traffic have been integrated into WoT infrastructure [35].

1.6.3 Cities

WoT technologies can be used for development of smart cities. Smart cities are an environment where every device is connected to the WoT network with minimal human interference. A smart city consists of smart homes, automation of traffic lights and smart parking. It should be able to reduce the stress in human life by automating every single task. A large-scale experiment in the city of Santander in Spain is described as a smart city experiment [36].

1.6.4 Industry

WoT technology that is used in the industry has been a topic of great interest [37]. A vast number of projects in the WoT sector have been conducted that are related to agriculture, food processing, security and surveillance. Until now, only the industries related to logistics, manufacturing and retailing have been attracted by WoT. With the advancement of wireless communication, smartphones and sensor technologies, smart objects are becoming involved in WoT.

1.6.5 Healthcare and Lifestyle

Recently, the convergence of WoT architecture made the development and dissemination of smart healthcare systems possible.

iHome has been proposed by the authors; this is a WoT platform for health which provides in-home services for healthcare on the basis of WoT and also integrating an intelligent medicine box (iMedBox) for assessment of different medical faculties with the use of sensors and devices [38]. Sebastian et al. [39] describe WoT architecture in different sports such as soccer in which healthcare is a priority concern.

1.7 Vulnerabilities, Attacks and Countermeasures

The advancement of technologies brings both positive and negative points. In one way, the technology brings additions to new features and makes human life comfortable, however, these new features are also vulnerable to attacks. The physical network and application layers of WoT can be susceptible to many attacks. This section will discuss these attacks in detail.

1.7.1 Physical Attacks

Physical attacks focus on hardware components that are present in WoT systems. These types of attacks affect the lifeline and functionality of the hardware. Different types of physical attacks are discussed further below.

1. Node Tampering

    This type of physical attack involves the damage that can be done to the sensor node by either changing the whole hardware part or by gaining access to that node and altering the sensitive data such as cryptographic keys [40].

2. RF Interference on RFIDs

    RFID tags can be easily hacked in order to send a noise signal created by the attacker. This noise signal is sent over the frequency of those radio signals that are used by the RFIDs for communication. This can easily result in a Denial of Service (DOS) attack [41].

3. WSN Node Jamming

    This attack is similar to the RF interference attack, except that this attack is WSN based. The radio frequencies which are used by WSN can be interfered with by the attacker which will result in signal jamming and communication denial to the nodes. If the jamming of key sensor node is successful, the attack can deny services for the WoT [42].

4. Sleep Deprivation Attack

    In the WoT systems, the sensor nodes are powered by replaceable batteries and are also programmed in way that they follow sleep routines so that battery life can be extended. Sleep Deprivation Attacks result in keeping the nodes alive, which increases power consumption and can ultimately result in the shutdown of the node.

5. Malicious Code Injection

    In this attack the node is compromised by injecting malicious code which results in providing the attacker with access to the WoT system. With the help of malicious code injection, full access to the node, or even the whole system, can easily fall into the hands of an attacker [43].

1.7.2 Network Attacks

These attacks are based on the WoT network layer and can also be executed remotely far from the WoT system over the internet.

1. RFID Spoofing

    For reading and recording of data transmission, RFID signals are spoofed. After spoofing of RFID signals, the attacker can transmit malicious data along with the original ID tag so that it appears to be valid [44].

2. RFID Unauthorized Access

    In RFID systems, as there are no proper authentication mechanisms present, anyone can gain access to the RFID tags. This obviously means that anyone, including an attacker, can manipulate the data by any means present on the RFID node [45].

3. Sinkhole Attack

    In a sinkhole attack, all the traffic that comes from the WSN nodes is lured away by the attacker resulting in the creation of a metaphorical sinkhole. In this attack, confidentiality of data is breached, and also service is denied to the network, which thus results in dropping of all the packets instead of forwarding them.

4. Denial of Service (DOS)

    In DOS, an attacker transmits more data to the WoT network than it can handle and as a result the WoT system starts denying service to genuine users.

5. Sybil Attack

    A Sybil node is known as a malicious node. It is a singular node that contains the ID of every other node. The WSN node accepts false information under this kind of attack.

1.7.3 Software Attacks

In any computerized system, the main cause of vulnerabilities is software attacks. Some of the software attacks that cause improper functioning of WoT devices are as follows:

1. Phishing Attacks

    In this attack, confidential information is gained by the attacker by using infected emails or phishing websites [46].

2. Viruses, Worms and Other Spyware

These are defined as malicious software that can affect the system and may result in various outcomes such as stealing and tampering with information, or even a denial of service attack.

3. Malicious Scripts

As WoT networks are always connected to the internet, the user that is controlling the gateway can be easily fooled in to executing ActiveX Scripts, which can ultimately result in system failure and complete shutdown of services.

1.8 Security Requirements

There is a continuous transfer of information between the WoT devices. Such an environment requires tight security for data protection. Security requirements of WoT systems for the protection of user data are: authentication and confidentiality and access control [47].

1.8.1 Authentication and Confidentiality

Authentication is defined as a process through which the credentials provided by a user or client are compared to what is present in the database for user authorization. In the context of authorization, the approach presented by Zhao [48] used a custom encapsulation mechanism that is a WoT security protocol. This protocol is called an intelligent service security application protocol, and it can perform critical operations such as merging of signature, encryption and authentication.

Kothmayr et al. [49] presented the first full implementation of a two-way scheme for authentication security for WoT which operates on the basis of current internet standards. Authentication for WoT has to be robust and highly automated. Access control helps for achieving confidentiality of user data by preventing unauthorized node use and by preventing the nodes being compromised. Confidentiality is defined as protecting the user's personal information when that information is shared over a network that is publicly accessible.

1.8.2 Privacy in WoT

WoT is now a part of various practical applications such as smart parking, traffic control, inventory management and more, and when using these applications, a user expects privacy of personal information. For management of privacy in WoT, data tagging was proposed. With the use of technologies that are present in Information Flow Control, data which is represented with the help of network events are tagged with different privacy properties. These tags help in the preservation of privacy by allowing system reasoning with the flow of data. Cao et al. [50] proposed Continuously Anonymizing STreaming data via adaptive cLustEring (CASTLE). CASTLE is based on a cluster approach that ensures constraints and anonymity for a data stream that results in enhancing preservation of privacy techniques (e.g. k-anonymity).

1.8.3 Trust in WoT

Trust is a notion of which there is no definitive consensus defined in the scientific literature, but which plays a vital role in WoT. Trust works strictly on the basis of management of identity and controlling access issues. The protocol of trust management for WoT is activity based, distributed and based on encounter as well. In this protocol, when two different nodes come in contact with each other, they can exchange trust evaluation about the other nodes. Thus, this type of dynamic trust protocol is able to adapt and choose from the best trust parameter and hence can adapt to the changing environment for achieving maximum application performance [51].

1.9 Relation of WoT and M2M

Machine-to-Machine (M2M) is a term in which a network is used by the machines for interconnection and these machines work without any user interaction. M2M is all about communication and connection with a "thing" that can be a machine, device or sensor or anything that can receive or send data. WoT is an elaboration of M2M. While M2M is all about connecting devices to each other, WoT is all about device interaction.

Connectivity is provided by M2M that offers capability to WoT. Without the concept of M2M, the concept of WoT must only be a pipedream [52]. M2M type of communication helps in the integration of different technologies that are required for communication in WoT. Services such as data transport and security are provided by an M2M service layer.

1.10 Conclusion

WoT (WoT) integrates different types of devices for serving different purpose such as sensing, processing, identification and communication. There is a rapid increase in WoT technologies as the sensors and actuators are getting more powerful and inexpensive with technical advancements. This chapter reviews recent research in WoT. First, the authors introduced the background and different types of WoT architectures. Next, we addressed the key technologies that play a vital role in WoT and protocols which are used in the WoT network. This chapter also discusses various application areas where WoT plays an important role presently and in the future. The major contribution of this chapter is the focus on the working of WoT technologies for future researchers.

References

1. Uckelmann, D., Harrison, M., & Michahelles, F. (2011). An architectural approach towards the future Internet of Things. In: *Architecting the Internet of Things* (pp. 1–24). Springer, Berlin, Heidelberg.
2. Huang, Y., & Li, G. (2010, August). Descriptive models for Internet of Things. In *2010 International Conference on Intelligent Control and Information Processing* (pp. 483–486). IEEE.
3. Atzori, L., Iera, A., & Morabito, G. (2010). The Internet of Things: A survey. *Computer Networks, 54*(15), 2787–2805.
4. Suo, H., Wan, J., Zou, C., & Liu, J. (2012, March). Security in the Internet of Things: A review. In *2012 International Conference on Computer Science and Electronics Engineering* (Vol. 3, pp. 648–651). IEEE.
5. Silva, B. N., Khan, M., & Han, K. (2018). Internet of Things: A comprehensive review of enabling technologies, architecture, and challenges. *IETE Technical Review, 35*(2), 205–220.
6. Bilal, M. (2017). A review of Internet of Things architecture, technologies and analysis smartphone-based attacks against 3D printers. *arXiv Preprint ArXiv:1708.04560.*
7. Khan, R., Khan, S. U., Zaheer, R., & Khan, S. (2012, December). Future internet: The Internet of Things architecture, possible applications and key challenges. In *2012 10th International Conference on Frontiers of Information Technology* (pp. 257–260). IEEE.
8. Papazoglou, M. P., & Van Den Heuvel, W. J. (2007). Service oriented architectures: Approaches, technologies and research issues. *The VLDB Journal, 16*(3), 389–415.
9. de Deugd, S., Carroll, R., Kelly, K., Millett, B., & Ricker, J. (2006). SODA: Service oriented device architecture. *IEEE Pervasive Computing, 5*(3), 94–96.
10. Buettner, M., Greenstein, B., Sample, A., Smith, J. R., & Wetherall, D. (2008, October). Revisiting smart dust with RFID sensor networks. In *Proceedings of the 7th ACM Workshop on Hot Topics in Networks (HotNets-VII).*
11. Floerkemeier, C., Roduner, C., & Lampe, M. (2007). RFID application development with the Accada middleware platform. *IEEE Systems Journal, 1*(2), 82–94.
12. Ray, P. P. (2018). A survey on Internet of Things architectures. *Journal of King Saud University-Computer and Information Sciences, 30*(3), 291–319.

13. Sun, C. (2012). Application of RFID technology for logistics on Internet of Things. *AASRI Procedia, 1*, 106–111.
14. Aggarwal, R., & Das, M. L. (2012, August). RFID security in the context of Internet of Things. In *Proceedings of the First International Conference on Security of Internet of Things* (pp. 51–56). ACM.
15. Madakam, S., Ramaswamy, R., & Tripathi, S. (2015). Internet of Things (IoT): A literature review. *Journal of Computer and Communications, 3*(5), 164.
16. Fan, Y. J., Yin, Y. H., Da Xu, L., Zeng, Y., & Wu, F. (2014). IoT-based smart rehabilitation system. *IEEE Transactions on Industrial Informatics, 10*(2), 1568–1577.
17. Yaacoub, E., Kadri, A., & Abu-Dayya, A. (2012, October). Cooperative wireless sensor networks for green Internet of Things. In *Proceedings of the 8h ACM Symposium on QoS and Security for Wireless and Mobile Networks* (pp. 79–80). ACM.
18. Akyildiz, I. F., Su, W., Sankarasubramaniam, Y., & Cayirci, E. (2002). Wireless sensor networks: A survey. *Computer Networks, 38*(4), 393–422.
19. IEEE 802.15, http://ieee802.org/15.
20. e–SENSE, http://www.nooviz.com/michel/eSENSE.
21. Arsénio, A., Serra, H., Francisco, R., Nabais, F., Andrade, J., & Serrano, E. (2014). Internet of intelligent things: Bringing artificial intelligence into things and communication networks. In: *Inter-Cooperative Collective Intelligence: Techniques and Applications* (pp. 1–37). Springer, Berlin, Heidelberg.
22. Ubi e–SENSE, https://ubisense.com.
23. Islam, M. M., Hung, P. P., Hossain, A. A., Aazam, M., Morales, M. A. G., Alsaffar, A. A., Lee, S.-J., & Huh, E. N. (2013). A framework of smart Internet of Things based cloud computing. *Research Notes in Information Science (RNIS), 14*, 646–651.
24. Rao, B. P., Saluia, P., Sharma, N., Mittal, A., & Sharma, S. V. (2012, December). Cloud computing for Internet of Things & sensing based applications. In *2012 Sixth International Conference on Sensing Technology (ICST)* (pp. 374–380). IEEE.
25. He, W., Yan, G., & Da Xu, L. (2014). Developing vehicular data cloud services in the IoT environment. *IEEE Transactions on Industrial Informatics, 10*(2), 1587–1595.
26. Jiang, L., Da Xu, L., Cai, H., Jiang, Z., Bu, F., & Xu, B. (2014). An IoT-oriented data storage framework in cloud computing platform. *IEEE Transactions on Industrial Informatics, 10*(2), 1443–1451.
27. Narendra, N., Ponnalagu, K., Ghose, A., & Tamilselvam, S. (2015, September). Goal-driven context-aware data filtering in IoT-based systems. In *2015 IEEE 18th International Conference on Intelligent Transportation Systems* (pp. 2172–2179). IEEE.
28. Bicknell (2009). *IPv6 Internet Broken, Verizon Route Prefix Length Policy.*
29. Shelby, Z., Hartke, K., Bormann, C., & Frank, B. (2013). "Constrained application protocol (CoAP).draft-ietf-core-coap-18," The Internet Engineering Task Force– IETF., Fielding, R. T. (2000). "Architectural styles and the design of network based software architectures," Diss., University of California.
30. Colitti, W., Steenhaut, K., De Caro, N., Buta, B., & Dobrota, V. (2011, October). Evaluation of constrained application protocol for wireless sensor networks. In *2011 18th IEEE Workshop on Local & Metropolitan Area Networks (LANMAN)* (pp. 1–6). IEEE.
31. Al-Fuqaha, A., Guizani, M., Mohammadi, M., Aledhari, M., & Ayyash, M. (2015). Internet of Things: A survey on enabling technologies, protocols, and applications. *IEEE Communications Surveys and Tutorials, 17*(4), 2347–2376.

32. Saint-Andre, P. (2011). Extensible messaging and presence protocol (XMPP): Core.
33. Standard, O. A. S. I. S. (2012). Oasis Advanced Message Queuing Protocol (AMQP) version 1.0. *International Journal of Aerospace Engineering. Hindawi www. hindawi.com*, 2018.
34. Heuser, L., Nochta, Z., & Trunk, N. C. (2008). *ICT Shaping the World: A Scientific View*.
35. Bandyopadhyay, D., & Sen, J. (2011). Internet of Things: Applications and challenges in technology and standardization. *Wireless Personal Communications*, 58(1), 49–69.
36. Sanchez, L., Muñoz, L., Galache, J. A., Sotres, P., Santana, J. R., Gutierrez, V., Ramdhany, R., Gluhak, A., Krco, S., Theodoridis, E., & Pfisterer, D. (2014). SmartSantander: IoT experimentation over a smart city testbed. *Computer Networks*, 61, 217–238.
37. Li, Y., Hou, M., Liu, H., & Liu, Y. (2012). Towards a theoretical framework of strategic decision, supporting capability and information sharing under the context of Internet of Things. *Information Technology and Management*, 13(4), 205–216.
38. Yang, G., Xie, L., Mäntysalo, M., Zhou, X., Pang, Z., Da Xu, L., Kao-Walter, S., Chen, Q., & Zheng, L. R. (2014). A health-IoT platform based on the integration of intelligent packaging, unobtrusive bio-sensor, and intelligent medicine box. *IEEE Transactions on Industrial Informatics*, 10(4), 2180–2191.
39. Sebastian, S., & Ray, P. P. (2015). When soccer gets connected to internet. *Proceedings of the I3CS*, Shillong (pp. 84–88).
40. Perrig, A., Stankovic, J., & Wagner, D. (2004). Security in wireless sensor networks.
41. Li, L. (2012, May). Study on security architecture in the Internet of Things. In: *Proceedings of the 2012 International Conference on Measurement, Information and Control* (Vol. 1, pp. 374–377). IEEE.
42. Mpitziopoulos, A., Gavalas, D., Konstantopoulos, C., & Pantziou, G. (2009). A survey on jamming attacks and countermeasures in WSNs. *IEEE Communications Surveys and Tutorials*, 11(4), 42–56.
43. Andrea, I., Chrysostomou, C., & Hadjichristofi, G. (2015, July). Internet of Things: Security vulnerabilities and challenges. In *2015 IEEE Symposium on Computers and Communication (ISCC)* (pp. 180–187). IEEE.
44. Mitrokotsa, A., Rieback, M. R., & Tanenbaum, A. S. (2010). Classification of RFID attacks. *Gen*, 15693, 14443.
45. Lin, J., Yu, W., Zhang, N., Yang, X., Zhang, H., & Zhao, W. (2017). A survey on Internet of Things: Architecture, enabling technologies, security and privacy, and applications. *IEEE Internet of Things Journal*, 4(5), 1125–1142.
46. Jagatic, T. N., Johnson, N. A., Jakobsson, M., & Menczer, F. (2007). Social phishing. *Communications of the ACM*, 50(10), 94–100.
47. Roman, R., Zhou, J., & Lopez, J. (2013). On the features and challenges of security and privacy in distributed Internet of Things. *Computer Networks*, 57(10), 2266–2279.
48. Zhao, Y. L. (2013). Research on data security technology in Internet of Things. In: *Applied Mechanics and Materials* (Vol. 433, pp. 1752–1755). Trans Tech Publications: Zurich, Switzerland.
49. Kothmayr, T., Schmitt, C., Hu, W., Brünig, M., & Carle, G. (2013). DTLS based security and two-way authentication for the Internet of Things. *Ad Hoc Networks*, 11(8), 2710–2723.

50. Cao, J., Carminati, B., Ferrari, E., & Tan, K. L. (2010). Castle: Continuously anonymizing data streams. *IEEE Transactions on Dependable and Secure Computing,
8*(3), 337–352.
51. Sicari, S., Rizzardi, A., Grieco, L. A., & Coen-Porisini, A. (2015). Security, privacy
and trust in Internet of Things: The road ahead. *Computer Networks, 76,* 146–164.
52. Severi, S., Sottile, F., Abreu, G., Pastrone, C., Spirito, M., & Berens, F. (2014, June).
M2M technologies: Enablers for a pervasive Internet of Things. In *2014 European
Conference on Networks and Communications (EuCNC)* (pp. 1–5). IEEE.

2

WoT-Enabled Delay-Tolerant Networks

Siddhant Banyal, Kartik Krishna Bhardwaj and Deepak Kumar Sharma

CONTENTS

2.1 Introduction

2.1.1 Contemporaneous Work and History of the Web of Things and Delay-Tolerant Networks (DTNs)

Delay-tolerant networks or DTNs are characterized by extended latency and unstable network topology where network contacts are intermittent, and there is an absence of a route between the source and goal node for most of the time. The Web of Things (WoT) is the physical realization of the internet web (World Wide Web) principle, where the software nodes of the internet are replaced by physical data nodes (such as people, things, sensors), with the uppermost network layer for the architecture being the internet itself. The data nodes used here are components associated with the Internet of Things (IoT) and hence it is safe to surmise that WoT is the implementation of IoT architecture fused with an internet service-based superseding network layer. In principle, DTN routing works in a two-phase process where the nodes store the message and forward the data package when it finds a suitable node to pass on the data as and when connectivity is restored. The following picture describes the general architecture for a DTN network (Figure 2.1).

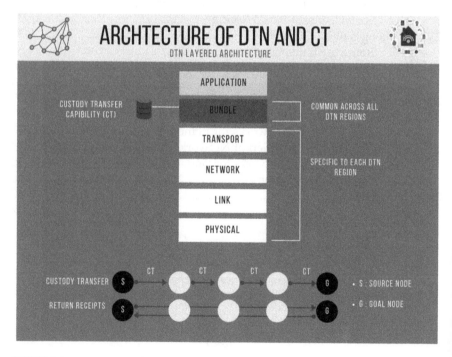

FIGURE 2.1
DTN-layered architecture and CT.

These features render the existing mobile ad hoc network protocol unsuitable because they assume the existence of a complete, two-directional and connected path in conjunction with a high-delivery ratio and link reliability. When combined with the WoT scheme, network nodes aim to make long interspace communication through the nearest node with internet access in the node network web, which shall forward the message to the required receiver or its nearest neighbor through already established internet web technology. Delay-tolerant networks are a reliable mode to enable data transmission where there is a lack of sophisticated infrastructure and the end-to-end path is absent at the moment for data transmission. This chapter highlights a myriad of applications of DTNs that experience the mobility issues of a lack of continuous links between the nodes and transient topology. This includes, but is not limited to, wildlife tracking networks, monitoring scientific and hazardous events and providing communication systems and data transfer facilities to remote rural sites, underwater networks and interplanetary networking (IPN). In addition, this chapter highlights a myriad of routing techniques for DTN that addresses the provision to transfer and direct bundles starting from the source node to the goal.

They are taxonomically assessed and categorized into four forwarding and flooding-based DTN routings:

1. Epidemic routing
2. Probability-based routing
3. Routing protocols based on geography
4. Social concept-based routing

In addition to the reliability of intermittent communication provided by DTNs toward a network of mobile and sparsely connected nodes, WoT adds the usability of an already established network into the DTN message parcel scheme, decreasing the inter-node infrastructure reliance and hence helping congestion in the local DTN network. It also allows the localized DTN network to communicate with nodes outside this network through the internet, increasing the information pool availability for such a network.

2.2 Routing Protocols in DTNs

2.2.1 Routing Considerations

As discussed in the previous section, delay-tolerant networks are a class of network such that there is an absence of complete connectivity in terms of a path. This area is currently and heavily being researched due to prospects of applications in terrestrial, space and other networks that are extremely challenged and require opportunistic routing paradigms. In contrast to traditional ad hoc networks, DTNs have specific challenges on the following fronts:

1. *Delivery*: Ratio, latency and cost
2. *Efficiency*: Data, coverage and security
3. *Resource*: Energy and overhead

The above challenges require specific considerations while developing routing solutions. In principle, DTN protocols work on a Store-Carry-Forward principle where if any message is dropped by any node in between the networks, it may be stored, then forwarded as and when an intermediate node appears. There exists a plethora of metric-based indicators to evaluate the efficacy of a routing protocol. They may be divided into three broad categories:

Delivery based:

1. *Delivery ratio*: Is the ratio of messages that are generated to messages that are actually delivered to the goal node.
2. *Delivery latency*: Latency evaluates the efficiency of the routing path chosen based on message generation and delivery.
3. *Delivery cost*: Evaluated using the magnitude of copies of the data produced by messages, is directly related to the overhead cost.

Efficiency based:

1. *Data effectiveness*: Ratio of data traffic produced to the unique multicast data packets delivered successfully to the users.
2. *Coverage*: Percentage of required goal nodes possessing a copy of message TTL (Time To Live), the expiration for the message.
3. *Security*: Based on privacy, anonymity and access control.

Resource based:

1. *Energy*: Supply of energy to the mobile nodes in network is limited. This metric depends on the number of copies and the number of messages delivered to the destination node.
2. *Overhead*: This encompasses the storage and bandwidth required for the network and is evaluated against successful delivery.

Developing any network routing protocol depends on the above performance metrics, such as reduction in transmission latency, reducing overhead, increasing delivery ratio, and so on (Figure 2.2).

The Web of Things (WoT) has attracted the attention and fascination of the research community [1]. The Internet of Things (IoT) has asserted itself as

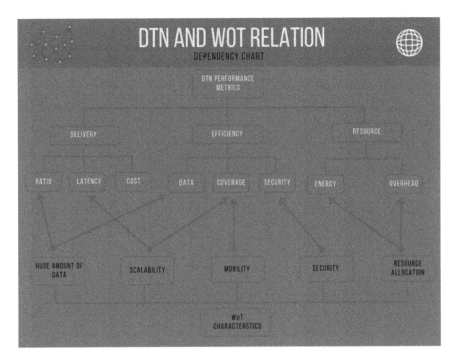

FIGURE 2.2
DTN and WoT – Relationship and dependency.

the engine that runs our modern technological endeavors under the ambit of public and private domains. Industries along with business institutions have evolved to be more reliant on IoT to perform essential missions and functions to improve productivity. As of now, there are over 23 billion IoT devices across the globe and this figure is expected to increase and reach the staggering amount of 60 billion in just half a decade. In conjunction with improving productivity, this has led to huge economic growth with the market size in Europe expected to shoot up to €242,222 million by the end of 2020 [2]. As of now, WoT has transcended and asserted itself as a pivotal communication media on the internet and has proven to be an indispensable compatible application in the current diverse infrastructure. As a result, there exists a tendency to use current technology and infrastructure to link the physical and cyber elements. Essentially, the Web of Things or WoT is a style of software architecture that allows elements of the physical world (objects) to be linked up to the World Wide Web.

2.2.2 Routing Protocol Classification

As discussed in Section 2.1.1, routing in DTN is a challenge due to the absence of an end-to-end path between the sender node and the goal node. In addition, the paucity of resources exacerbates this challenge. Here, the advantage of node's mobility is taken into consideration when making routing considerations. In these challenged networks, a progressive step-based approach is

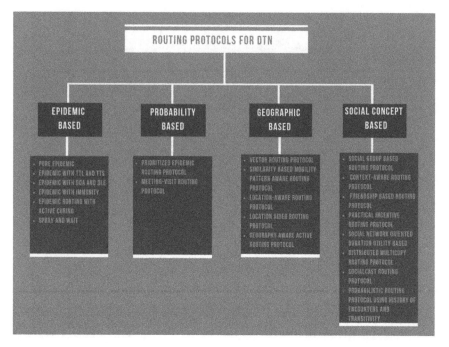

FIGURE 2.3
Routing protocol classification.

taken where routing comprises local forwarding considerations. Some routing considerations assume there is no information or prior knowledge of the network, while some factor in certain qualitative indicators to exploit that information for routing consideration. This section aims to explore different kinds of routing protocol classes and aims to build a base for the various protocols that we will discuss in further sections.

The main aim of current DTN routing protocols is to augment the probability of discovering a path in an environment where there is paucity of information about the network. This includes a myriad of mechanisms, like stationary waypoint stores, replication of messages, approximating encounter probabilities, network coding and leveraging prior knowledge, and so on. For this section, we shall consider some key protocols while discussing the various classes of routing protocols from Figure 2.3.

2.3 Forwarding and Flooding-Based DTN Routing

2.3.1 Epidemic-Based Routing Protocols

Epidemic routing stems from the principle of epidemic algorithm which was proposed by Vahdat and Becker [3]. This protocol works on the minimum

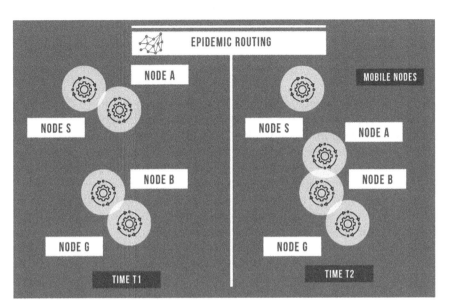

FIGURE 2.4
Epidemic routing.

assumption of information pertaining to networks such as topological information and network connectivity. The delivery of messages happens in a pairwise manner so as to spread information and finally reach the goal mode. Through this method, messages travel quickly in segments that are linked; further, this method has a high probability of reaching the goal node. We can better understand this through an example, as illustrated in Figure 2.4. Assume there is a node S (Source) that aims to send a message to a destination or goal node G (Goal), but there is no direct path. S has node A in its neighborhood, and node B is in the neighborhood of G. In this type of routing, since the path to the goal node is not available, messages are stored in buffers at the various hosts and a table of summary vectors is maintained at each host. These summary vectors collect their own information and exchange this data when connected to other nodes. After this exchange, the nodes determine whether other nodes have messages they did not receive before; if they don't, summary vectors are updated so as to encompass the new messages. There are two key requirements for this type of routing: First, a large buffer space so that each node has the capacity to hold messages in an epidemic manner, and second, the presence of global IDs for messages in order to make the determination whether they received the message or not, as discussed. Although this has a higher delivery ratio success, it uses a large buffer and network resources which may be impractical in a constrained environment. For this we use metrics such as Time To Live and Time To Send (TTS) for a more practical realization and routing of messages.

TTL and TTS: Time To Live, abbreviated as TTL, limits the lifetime of a data in a system or a network and is often called a hop limit. It is implemented as a frequency counter or a clock associated with the data packet. Time To Send works on a similar principle, taking the time required to send the message as a parameter.

DOA: Drop Oldest, abbreviated as DOA, is a routing strategy that aims to drop the oldest packet in the network. This works on the rationale that if the packet is the oldest in the network, it is likely to have been delivered.

DLE: Drop Least Encountered, abbreviated as DLE, is another routing strategy that is based on delivery ability, which is determined based on history and analysis. Comparative studies show DLE is more effective than DOA.

Another popular method under the epidemic routing is the Epidemic with Immunity list [4]. In this, certain intermediary nodes are selected based on popularity (likelihood) so as to increase the delivery ratio at the cost of higher end-to-end delay. Each node maintains an Immunity list which they exchange upon encounter so as to avoid duplication and forwarding. This is similar to the summary vector list, with the addition of the IDs of messages that have been delivered. Mundur, Seligman and Lee were the first to propose this in 2008. Spyropoulos et al. [5] put forth the Spray and Wait routing mechanism with an Epidemic Protocol so as to leap ahead in delivering messages. Spray and Wait encompasses two phases. In the first phase, K copies of the messages are spread across the network to K relay nodes; this is called the spray phase. The next phase is called the wait phase. If the message does not reach the goal node, each of the K distinct relay node performs direct transmission via unicast to transmit the message. This method is essentially a trade-off between the unicast and multicast transmission. Binary Spray and Wait is a variation of the above algorithm proposed by Spyropoulos. In this method, K/2 copies of the messages are "sprayed" in the network, where K is the total number of copies, with the node keeping the rest [K/2] of the copies itself. This process of spraying is done until there is just one copy of a message left with the node post while the wait step is implemented. It has been noted that this method has a higher rate of delivery and faster transmission than the conventional Spray and Wait method proposed.

2.3.2 Probability-Based Routing

This type of routing is quite prevalent in opportunistic environments and deployment to these strategies in DTN routing protocols. The principal idea behind this sort of routing is using probabilities to predict the information regarding the network at a time instance in the future. This may employ parameters like message delivery probability and the next time of encounter for the nodes based on the history of encounter for the nodes. Lindgren et al. [6] proposed the PER protocol that worked on the principle of probabilistic routing which is based on a time-homogeneous model which is a Markov renewal process with the state defined for every state, including the jump states. While determining the transfer, the PER model uses three mathematical parameters, out of which one is used by the nodes to ascertain the relay nodes. The results show this routing mechanism improves delivery ratio and reduces latency relative to conventional DTN routing schemes.

$$\alpha = \mathrm{Max}\, C_{na}(k), 1 < k \le D \quad \max \quad \beta = \sum_{k=1}^{D} C_{nad}(k) \quad \gamma = \sum_{k=1}^{D} R_{nad}(k)$$

where:
k = Discrete time slot
D = Maximum message acceptable delivery delay
C_{na}: Probability of na (the selected neighborhood of node a) and destination d connecting at t
R_{nad}: Probability of first connection of na and d at time t

Prioritized Epidemic Routing (PREP) proposed the idea of dropping nodes based on cost from source, the cost to destination, the packet expiry time and the packet generation time to save resource usage without impeding the delivery ratio in the network. It was proposed by Ramanathan et al. [7] in 2007 with the idea of utilizing the storage and bandwidth by dropping only when required so as to increase efficiency.
The two modular of PREP encompasses:

• Independent components: Calculate the cost of routing from a specific node to the goal node based on topology at that time.
• Priority scheme: Encompasses transmitting when there is no transmission for t seconds and the deleting process which initiates when the usage of buffer is past the set threshold value.

Meeting Visit Routing Protocol (MV) assesses the history of information like the frequency and time of visit at a specific location. The selection basis is dependent on the probability of delivery $P^{an}(i)$, that estimates the delivery probability form the current node "a" to the goal node "i" with n jumps. The formula for delivery probability is defined as:

$$P^{an}(i) = 1 - \prod\left(1 - m^{ja} P^{jn} - 1(i)\right),$$

M^{ja}: Probability of meeting node "j" and node "a" within last t cycle.

Upon encountering another node, "a" exchanges the message, and along with that updates the delivery probability, and the list gets sorted based on the P node, "a" eliminates its own message based on a lower probability index in reference to its neighbors and selects the best n messages (based on P) to forward and receive messages from its neighbor that it does not have.

2.3.3 Routing Protocols Based on Geography

Routing based on geography requires information about the movement pattern of DTN nodes and this can reduce and make up for the rugged landscape by enabling fixed mobile nodes. Vector Routing Protocol (VCR) was first proposed by Kim and Kang [8], which employed the node's information pertaining to the location so as to compute the velocity and direction to the destination. This methodology reduces the buffer usage without impeding the delivery ratio. The node that is mobile employs the Vector V^{cur} along with the pattern based on movement history so as to estimate a future vector. While considering this form of vector routing, nodes which are traversing in a direction that is orthogonal exhibit higher frequency of packet replication in relation to nodes that are linearly traversing (same or opposite direction). In addition, the magnitude of velocity is also considered since it is more sensible to forward the data packet to the faster node rather copying to the slower node. Simulation model results show that this has a comparable delivery ratio compared to the epidemic routing discussed in Section 2.3.1, but shows reduced traffic in comparison. Yin and Cao [9] proposed the similarity-based mobility pattern that employs GPS data and compute their mobility pattern to route data packets. The DTN pattern in this route in this work considers the secondary node's velocity and the goal node (node g) with the angle (θrg). Depending on the information pertaining to the location of the node $f(r,g)$, a similarity index is calculated to elect the next forwarding node in the neighborhood by one hop.

$$f(r,g) = \lambda 1 \times 1 / \theta rg + \lambda 2 \times \left(\left(xr - xg \right) 2 + \left(yr - yg \right) \right)^{-0.5}$$

The Location Aware Routing Protocol considers the issue of isolation in a delay-tolerant network and considers both the information pertaining to location and the social network structure while making routing considerations. In a case where the given node and the goal node are in the same component and the node which the given node met has a parameter of a higher value, then the data packet is copied to the node which the given node just met. If not, the given node waits for nodes that are available later. The operation is bifurcated into forward mode and replicate mode such that a node may forward only one post in which the node becomes inactive, whereas it may replicate nodes in absence of restriction. So a node may forward the node to the node it just met once, whereas if that node seems unfavorable, it copies the message. LAR or Location Aided Routing also exploits the data related to the location to be efficient in terms of overhead usage by restricting the search space for specific routes. Conventionally, in the route discovery phase, a node broadcasts its route request via flooding and upon encountering this request, the secondary node checks the goal node or destination and if they align, the node reaches the goal and if not, the secondary node forwards the data packet again.

Two functional units are defined in LAR:

- Expected area: Where the node has made a prior visit in order to ignore redundant nodes.
- Request area: Defines the area for route request (RR), including the expected area and its boundary.

Expected area can be thought of a radial unit, with the center of the circle being previously known, and the location of the goal node and radius being the product of speed of goal node (average speed) and the time period.

GAARP or Geography Aware Active Routing Protocol was first proposed by Fan et al. [10] so as to assess mobile DTN from a geographical perspective. This research defines a society-based concept that clusters under closely connected/linked items, like a geographical community, by accounting for geography-related social networks. The principal operation of this protocol depends on the concept of a centrality index that encompasses the density of the so-called "geo-community," which is dynamic in a method that is defined mathematically in the following manner:

$$P_i(S_i) = 1 - (1/N) \sum_{a=1}^{N} \left(1 - \phi_i^a\right)^{S_i}$$

- P_i: Average probability a user visits the geo-community and time S_i.
- Φ_i^k: Computes the steady state probability for node a in its ith geo-community.

2.3.4 Social Concept-Based Routing

The research community has been analyzing the mobility patterns to extract similar features with social network based on identifiable social relationship and communication. The qualitative metrics used to make decisions pertinent to routing and protocols based on these social models have been presented. Social group-based routing protocol (SGBR), proposed by Yin et al. [11], uses multicast routing while exploiting clustering based on social groups amongst nodes so as to deliver better performances in terms of delivery rations without bolstering network traffic. The characteristics of nodes from the same group have similar social behavior and are likely to meet frequently, hence a node from a social group is an "agent" of the group in consideration. The authors [11] proved that the probability of replicating a message in another social group is higher than within its own social group. The strength of the node is measured in terms of connectivity Δ_{xy} that quantifies frequency of encounter between x and y. Δ_{xy} gets updated by the following equation, such that:

$$\Delta_{xy} = \Delta_{xy}^{old}\beta^k + \left(1 - \Delta_{xy}^{old}\beta^k\right)\zeta$$

ζ: Updating factor, $\zeta \in (0,1]$
β: Aging constant, β y$\in [0, 1]$
k: Time elapsed since the nodes encountered

Exploring contextual information is another method to use to compute and estimate delivery probability which protocols like Context Aware Routing (CAR) aim to utilize. CAR considers mobility metrics, amount of battery and link-loss ratio. To enable this sort of computation, a weight method is used to maximize the value of α_i (α_i utility function $\rightarrow$ attribute a_i) for each attribute as per the equation below:

$$Maximize\ f\left(\alpha\left(a_i\right)\right) = \sum W_i a_i$$

such that,

W_i: Weight based on significance of each attribute
α_i: Utility function
a_i: Attribute

The friendship-dependent routing protocol proposed by Bulut et al. is based on three friendship behavioral metrics based on the history of encounters of the nodes:

- Frequency
- Longevity
- Regularity

In addition, for nodes that are directly connected, they use Social Pressure Metric, or SPM, to compute social pressure amongst nodes x and y to meet each other where j(t) is the time left before the other two nodes meet.
Friendship $\propto$ SPM^{-1}

$$SPM_{xy} = T^{-1}\int_{t=0}^{T} j(t)\,dt$$

The SocialCast protocol works on the principle of subscription where the messages are sent based on a subscription such that messages are only sent when there exists the same subscribed interest. SocialCast routing protocol has the following four phases:

- *Interest dissemination*: Broadcasting requests and exchange of the summary vectors that contains the relevant metrics.

- *Carrier selection*: Based on the highest utility among the neighbors, a node is selected to be the carrier.
- *Message dissemination*: The content reevaluation takes place and the message is forwarded to the best carrier.
- *Message publishing*: After the publishing phases, the published message is stored in the local buffer.

Probabilistic Routing Protocol using History of Encounters and Transitivity (PRoPHET) employs historical data to make probabilistic determinations that assist in making routing considerations for the nodes. The underlying principle for PRoPHET is similar to other routing protocols discussed in this section in that a node is likely to revisit its past location based on mobility pattern.

Delivery predictability is defined as $U_{(a,b)} \in [0, 1]$ such that "a" is mapped to "b" node. The value of delivery predictability is performed in three steps with the first step being to update the metric when nodes meet:

Where $U_{init} \in (0, 1]$ is an initialization constant.

$$U_{(a,b)} = U_{(a,b)\text{old}} + \left(1 - U_{(a,b)\text{old}}\right),$$

where $U_{init} \in (0, 1]$ is a constant for the initial value of U.

If a pair of nodes do not encounter each other often for a while, this implies that they may be decent forwarding agents with respect to each other, hence, this should reflect on the delivery predictability value and it must age accordingly.

$$U_{(a,b)} = U_{(a,b)\text{old}} \times \gamma_i$$

where Aging Constant $\gamma \in (0, 1)$ and i: Time interval since aging on metrics were implemented.

Further, the metric of delivery predictability is transitive in nature, if a node "a" encounters node "b" in a small-time interval with high frequency and node "b" bears the similar high meeting proclivity for node "c," then delivery predictability for a→c will bear a similar relation, hence, a is a good node to send a message to node c. This is property of transitivity illustrated in Figure 2.5 for examination.

$$U_{(a,c)} = U_{(a,b)\text{old}} + \left(1 - U_{(a,c)\text{old}}\right) \times U_{(a,b)} \times U_{(b,c)} \times \beta$$

where Scaling Constant: $\beta \in [0, 1]$, determines the scale of impact.

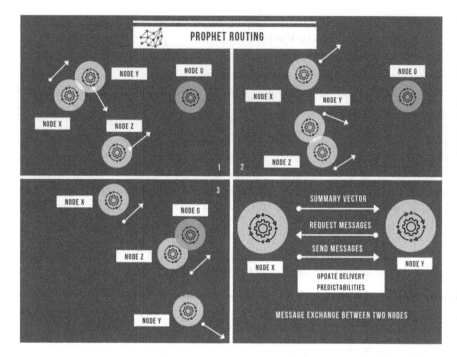

FIGURE 2.5
Transitive communication: A message is passed from node X to node G via nodes Y and Z through the mobility of nodes.

2.4 Challenges in WoT-Based DTNs

2.4.1 Routing Challenges

The delay-tolerant network, as the name suggests, is prone to delays in data delivery. Propagation delays are fairly acceptable in DTN schemes, and are known to occur variably with different data load scenarios, increasing with the size of data and decreasing with an improvement in node mobility of the network. With the delivery ratio highly dependent on the motion of nodes and their interactions with one another, there is a good chance that the DTN will give out a poor delivery-time rate, with the node interactions being random, and a high probability of node isolation and data coagulation [12]. This shall also lead to other issues like rapid energy drain due to constant search for a neighboring node for data transmission, as well as node buffer due to no neighboring node to further offload the data burden on the node.

Also, with a latency in the data delivery and a probability of data stagnation in the DTN architecture, there arises the possibility of poor data delivery ratio, with the possibility of data bundles being aged out by individual nodes due to maintenance of individual node efficiency factors employed in the DTN schemes.

The DTN Bundle Protocol has been improved over time to improve the data propagation rate as well as the data delivery ratio though largely leaving the application requirements due to a varied infrastructure usage over various applications of DTN, and hence not helping in improvements in application-based DTN architecture, even if theoretical improvements have been proposed [12].

To solve the above issues, many solutions have been proposed over time with each having their own implementation complications that the system might face during deployment.

To improve the node interactions and make the data exchange more intuitive, hence improving the delivery rate, opportunistic contact between nodes has been proposed [13]. Opportunism in networking based on the node activity and movement allows an optimized data forwarding outlet that improves the data delivery ratio as well as reducing propagation delays, while reducing the number of data hops, bringing the best properties of DTN and ad hoc networking together, hence providing the perfect amalgamation of the two. The downside of opportunistic networks include assumptions of node mobility patterns and data hop decision tables as well as the computational and storage requirements for the opportunistic operations. The assumptions made shall not hold good in every situation and in cases like node failure and out-of-pattern node movements, and the delivery ratio shall be lower than projected. Also, the complexity of operations for opportunistic network tabulations shall require a costlier hardware to support, with an imminent apprehension of overloading the computational systems and buffer memory overrun, impugning the efficiency of the said system. The scheme will also demand a heavier battery draw for operations per node, depreciating the power efficiency of the system.

Schemes toward feedback path maintenance have been proposed, borrowing from opportunistic networks, where data delivery feedback will be sent back to the sender node, which, if not received within a threshold time period, shall lead to the node multicasting the message to other nodes known to be able to carry data between the given nodes [13]. Such a scheme will increase the data delivery ratio without much computational requirements due to absence of involvement of node intelligence. However, such a system shall have an increased data storage requirement, also increasing the power requirements for the said operations. Also, despite an increase in data delivery ratio, a high probability of increase in delivery time might be inevitable, due to capricious delivery status in the first attempt.

Despite the randomness in node movements, the mobility of nodes is often describable, by reasonable approximations, with the natural movements seen in the wild. Such a pattern study shall be beneficial in judging the capability of a node's decision on the next hop for the data, depending on the current network topology and the previous changes in the same attempt to reach such a topology, allowing it to model the node movements. Such a model shall improve hop decisions, reducing the delivery time as well as increasing data delivery ratio [14]. However, just like with the opportunistic networking, an increased complexity in networking scheme will require higher computational and storage preconditions, which will also lead to decreased power efficiency for the operational system.

There are many other implementation variations in the DTN architecture and operational schemes, each suited for different applications that the DTN serves. Given the applicability, the different DTN modifications need to be evaluated according to the needs and requirements of the said implementation.

Though the variations and modifications offered in DTN give specifically focused network properties for the various applications in which the DTN design can be implemented, a number of these variations have their own downside. Given that the modifications are data delivery parameter based, the application basis of such modifications is not taken care of, and instead a specialized system needs to be designed for the applicable implementation of such variations [14]. Also, with heterogeneous topology and architecture for different variations of DTN, scalability of such networks might be limited, and expansion and consolidation of different network implementations, despite being in the practical vicinity to carry it out, might not be possible due to the architectural incompatibility. Hence, a need for a common basic architectural layout for all such variations is needed to be achieved, standardizing the basic protocol operations, keeping in mind the network scalability.

2.4.2 Custody and Congestion-Related Issues

Given the intermittent contact between nodes and an ever-changing network topology, storage and forwarding is an important technique in the DTN scheme. Data must be stored until the suitable next hop is not presented to the node in question for the bundle transmission. Indeed, even after transmission of said data bundle, the data still needs to be stored by the node after the transmission to the nest hop so as to ensure the transmission of the data further and handle crashes as well as data bundle dropping, ensuring transmission to the intended node. Needless to say, data buffer for each source is an important resource to such a network and might greatly affect the performance for data forwarding and propagation time for individual nodes involved in a particular data transfer convoy. Hence, storage and buffering are a decisive factor toward the efficiency of DTN network.

In a DTN network, buffering is used for mobile information exchange. In this process, a node might end up being on the receiving end of transmission of various messages in a data exchange convoy [15]. This might lead to the overflow of the buffer capacity, leading to failure of that node, hence leading to the dropping of bundle, which might affect both, the delivery ratio as well as the propagation time.

Slow interfacing by the node might also lead to a buffer overflow for such a node, with the rate of data influx being higher than the rate of data out flux.

With a varying topology, there is a higher probability of data collision followed by the buffer jamming due to contention between simultaneously transmitting nodes to a particular node. Occupation and transmission queues can be maintained so as to solve the contention and overflow, with this having its own issues of an increased power consumption as well as computational requirements for each node.

The array of high complexity DTN variations that might improve the delivery ratio and propagation time often need higher computational requirements for the smart operations they employ. The memory requirements for such computational systems might coagulate the storage capacity of individual nodes. An intuitive solution might be usage of higher memory systems for individual nodes, but such an implementation might be costly and the operations for a bigger memory might also require and draw more power, leading to a decreased power efficiency. A solution to such a problem is the employment of distributed computational solutions for DTN architecture designed to fit the requirements of its application, which would not only be efficient in computational transactions, but would also lead to a higher competence in the use of the data storage.

Communication networks might encounter blockages due to the resource overflow as well as network congestion, which would lead to delays, while simultaneously leading to an increase in the power consumption of the communication system. Solutions to such issues are proposed as follows.

For back-off mechanisms of resource allocation, employment of extended access barring (EAB), using an extended wait timer and a delay-tolerant indicator, is done so as to create a differential window for different sized packets instead of slicing of data, control on parameters of access delay and access probability, hence providing energy efficiency. Extending the EAB to four paging cycles further shows improved efficiency in energy usage as well as resource distribution [16].

Resource allocation to a modulation and coding scheme (MCS) is the key factor to energy efficient communications. It inhibits congestion and implements overload control due to intrinsic properties of choosing optimal MCS. The association of picking transmit power with MCS determination adds to the energy efficiency brought in by this method.

Clustering of devices, with treatment of each as a mobile node, and hence opportunistic usage of available memory related to each as a temporary buffer decentralizes the load on the central buffer and communication network, bringing down the network congestion in the process. It also saves energy by curtailing the need for transmission of data to the central exchange. Reinforcement of learning algorithms for selection of an enabled node-B (eNB), with access probability as its metric for choice, will help in distributed and dynamic networking, with median mobile nodes acting as a forward feed buffer, localizing and sectionalizing communication, and hence decreasing the overall congestion. Other methods like avoidance of near-simultaneous communication requests with larger back-off values to give a relatively spacious network time slicing and reference signal blocking based on statistical methods are also proposed, with simple yet effective implementations in employed protocols.

2.4.3 Security Impediments

Security of data is an important issue to be taken care of in any scheme of data networking. Preservation of information carried in the data exchange is a primary property that must be taken care of in any type of communication system. With the applications of DTNs ranging from satellite communications to military applications, the security concerns for DTNs are raised to higher stakes.

Some common issues that DTN communication network may face are listed in the following.

Vulnerability of the data center is a major security issue, where services and storage might be at risk of exploitation, disrupting the operations of the network.

Data encryption, which is a good tool for data protection, but needs careful evaluation in terms of implementations and choice of the protocol to be followed, can be a beneficial tool for information safety, though without a careful examination of the implementation schemes, the encryption might be easily breachable, leading to an easy leak of information through simple data trapping techniques. Also, tradition encryption schemes assume a continuous data flow through a network that is certainly not true for a storage and carry scheme-based DTN network, and therefore there is a need to modify the traditional encryption schemes to be implementable on the DTN-based solutions and applications. Identity-based encryption schemes are an important addition to encryption schemes for the DTN network, laying the basis for encryption in such disruption-based network topology [17].

Disruption of service by buffer flooding as well as demand spikes will lead to network crashing. This may be taken care of by maintenance of buffer stacks, so as to control overflow and resist flooding.

With the topology having various nodes that perform different functions, provision of access to the whole network topology as well as all the exchanged information might not be a good idea. Implementation of security control access levels to different nodes making up the DTN network according to their roles is a beneficial feature in terms of data and network security.

Given the nodes use the next hop procedure to forward the information, there is a possibility of passing the data to a malicious node created by perpetrators with intentions to intercept data. This is known as node spoofing and is a menacing security threat. Use of encryption will help secure the information in the event of data theft. In addition, the hopping protocols can be pre-empted by security checks between the transmitting and receiving nodes, leading to a secure gateway of information between the transmitting and receiving nodes, preventing node spoofing events. However, given the mobility of nodes, the window for information exchange might be small, and hence, such a call-receive security check system might not always be beneficial, leading to a trade-off between security and data delivery. Such situations can be prevented by the maintenance of node tables that shall be able to instantaneously verify the authenticity of the receiving node, though at a cost of extra memory capacities that must be added into the network capabilities [18]. Masquerading as authentic users and accessors of node-to-node hop-based architecture is an easier task due to lack of centralization and fixed path in DTN network topologies. This may lead to leakage of information during transmission and hence loss of information. This can be solved by usage of user registration systems with the use of personal keys as well as system-based authentication, like biometrics,

This will provide protection against compromising information through an unauthorized access leading to an active or a passive adversary of the DTN network.

2.5 WoT-Enabled DTN-Based Applications

2.5.1 Delay-Tolerant Networks for Satellite Communications

Currently, there are around 4,994 satellites orbiting in space. With the various applications for these orbiting satellites, communication with the satellites plays a vital role in bringing them to our use. Delay/Disruption Tolerant Networking (DTN) principles are primarily implemented in satellite-based network communications to help with this.

DTN is now extensively used in GEO (Geosynchronous Earth Orbit) satellite systems, replacing the conventional use of PEPs (Performance Enhancing Proxies).

Initially, DTN was developed for deep space communications and sensor networks. However, its application was later extended to satellite environments in order to get through the sporadic channels that are characteristic of LEO (Low Earth Orbit) constellation satellite systems. Since this network could handle disruptive channels, intermittent connections and the absence of an end-to-end connectivity of singular LEO satellites, as well as incomplete constellations, it is particularly useful in LEO systems [19].

With the growing applications for delay-tolerant networks, an inclination toward multi-satellite distributed systems is arising. Such systems comprise multiple small satellites interacting with each other so as to achieve distributed mission objectives. This further promises development in a new field which is merely a fusion of the conventional large monolithic satellites and small distributed group of satellite systems. For the realization of such systems, communication is a fundamental element where interaction between satellites is required [20]. Thus, prospective techniques in the field of satellite communication need to be explored.

For the realization of a distributed network architecture as proposed, a key aspect arises in the form of inter-satellite communication, that would form the backbone to such a proposition, establishing the need for an investigation toward the possible approaches for it.

A channel of efficient data exchange needs to be established while simultaneously catering to the challenges that pertain to networks of DSS (distributed satellite systems) in the earth's orbit. Given the handling of disruptive as well as delay-prone communication scenarios by DTN, the proposal of applications of DTN in DSS communication systems is promising, with DTN being capable of handling the challenges that arise in the DSS networks.

To understand the application of DTNs in satellite communication, we shall take into consideration the exchange of data between the ground station and a low-flying small satellite. It will be difficult for transfer of the complete data file in a single transaction with a particular ground station. The file might be too large for the same, given the time period of the existence of such a link. Hence, such a file should be divided into partitions, that will make up the different data bundles. These bundles, as many as possible, are then transferred to the ground station with which the satellite is currently linked. The other fragments of the files will be transferred to other ground stations as they are encountered. Each ground station then transmits the received bundles to a central station where all such bundles are compiled to get a reassembled file, the same as the one transmitted. The disruptions in terms of various factors such as climate, ground conditions and space disturbances will lead to inaccuracies in the information transmitted. The ground station will be responsible for identification and rectification of such errors. In such a case, a request for retransmission of the corrupt bundle will be made by the next ground station that satellite should be in contact with. The satellite will drop the data bundle stored during the next link with the subsequent ground station [21].

A visual explanation of the procedure is provided in Figure 2.6.

2.5.2 Delay-Tolerant Networks for Remote Communications

Communications have grown to become an indispensable resource in contemporary lives. That said, there are still unconnected areas with sparse networking infrastructures to support a modern communication system.

The requirements for setting up a communication scheme in such conditions are different from the normal end-to-end communication connections. The end-to-end connection will drop a data packet if not delivered by the time threshold. This is not the case with DTN network, which will assume

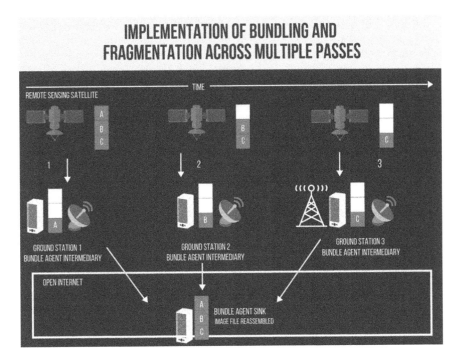

FIGURE 2.6
A visual explanation of DTN satellite communication, explaining data transmission between the ground stations and satellite and the compilation and information extraction from the received file.

that the destination path is unreachable. Hence, each node in a DTN communication network stores the data in its custody as long as the communication to the end node is not confirmed for the data bundle to be transmitted, ensuring the delivery of information to the destination node, guaranteeing that the data is not lost in the intermediate path to the destination.

The requirements for a remote communication are discussed below, explaining how DTN can be efficient in delivering within such requirements.

For every communication network, the cost effectiveness of the solution toward the communication problem is the basic requirement. Traditional communication channels assume usability over a wide base of population, which will not hold true for remote areas that are often sparsely populated. Hence, it will prove to be expensive to lay out such plans for them. With the advent of mobile communication, the employment of DTN networks, which rely on cheap connectivity between nodular topology, promises access to communication in a cost-effective manner. This lays out the background for success of deployment of DTN communication schemes for remote areas [22].

The user expects reliability of delivery from any communication setup, although with understandable delays. Such an expectation shall tolerate periodic connectivity over data loss–prone communication, even if it promises a constant connectivity. As described before, the mechanism of DTN communication deals with the issue of reliability in communication by using node-based storage for data propagation, which gives it an edge over end-to-end communication structure, hence being preferable in remote communication, providing a near-perfect guarantee of data delivery.

Efficiency in communication is another user expectation. With end-to-end communication being a costly affair for remote connectivity, the DTN, despite the delays, is a preferable choice for such communication scenarios, salvaging the efficiency of network. The DTN network's efficiency can be improved through various tweaks in DTN implementations for such scenarios.

A user is likely to use a wide range of devices while availing a communication scheme. Compatibility of a network with all such devices is an important requirement out of any communication scheme for it to be feasible to use. DTN implementations for remote communication take care of such requirements by adaption of properties from continuous connection-based networks, applying them into usability scheme with an opportunistic exchange between the communication channels and the end user, by making the network architecture more heterogeneous in nature [22].

The user's mobility needs to be taken into consideration for a communication platform, with usage of hand-held devices being the way of life in the modern times. Such requirements are served by the use of service node zoning and multiple transmission and receiving channels deployment, which is already an intrinsic property of DTN-based communication operations.

Security is an important factor of consideration and a general expectation of the user. A secure gateway of information flow is an important aspect for building a communication network. The DTN is complemented by various security measures, implementing both traditional and more DTN architecture and working suited methodologies, ensuring the user for information safety.

The world is predominantly connected by the end-to-end, traditional communication systems. It is, hence, a necessary requirement to ensure the capacity of DTN networks to be able to communicate with the continuous connection-based communication. Compatibility between the traditional and DTN-based remote communication solutions is vital to the network setup to avoid extra hardware implementations for realization of remote communication at a cheaper and affordable cost. The DTN network is already equipped for such adaptations, and hence is implementable [23].

Scalability is a desirable feature for user-based communication, allowing an increase in the scope of reachability of network for future expansions. The exploratory nature of DTN networking implementations, employing algorithms that continuously update the network topology, finds a perfect fit toward fulfillment of such requirements.

Some key features of a DTN-based communication network, that make it suitable for remote communication, are listed below [24]:

• Ability to ensure and take responsibility for the data bundle to reach the required destination of the network, fulfilled at a nodular level.

• Ability to cope with a disruptive connection between nodes in the network topology, if such a situation arises.

• Ability to cope with long delays in propagation and delivery of data bundles.

• Ability to make smart use of knowledge of scheduling, prediction and opportunism in connectivity (as an added advantage toward a continuous network connectivity).

2.5.3 Vehicular Delay-Tolerant Networks (VDTNs)

Intelligent transport system solutions are on the rise in the contemporary era. The introduction of intelligence in transportation, increasing safety of travel as well as providing more information about road parameters to the passenger, has had a marked improvement on the user's travel experience. However, such implementations have an inherent need for a robust mobile communication that holds an intrinsic property of a very dynamically evolving topology in the communication [25].

DTN-based communication has found its applications in such paradigms, suiting the needs of the given communication network. The DTN-based communication systems cater to the vehicular intelligent systems that require communication between the roadside smart architecture installations as well as vehicle-to-vehicle communication, which is characterized by frequent disruption as well as disconnectivity at times for understandably long periods. The scenarios characterized by a high delay in propagation and frequent partitioning are also a very common occurrence in such networks. In the past, solutions known as vehicular ad hoc network (VANET) have been proposed for realization of such a communication scheme [26].

With an improvement in VANETs in mind, DTNs have long been proposed for such applications, given their ability to handle the issues of vehicular networking, as stated above. Given the matching characteristics of DTNs to the given application, a new communication paradigm for this has been introduced, namely the vehicular delay-tolerant network (VDTN). The VDTN operates on the familiar store and carry forward mechanism, characteristic of DTNs, making it possible to realize an application having an inherent delay-tolerant nature without any end-to-end connectivity [27].

VDTNs are outlined with a relative short contact between nodes, given the dynamic nature of traffic, leading to a very unstable topology, where routing of data arises as challenging issue. A typical DTN approach of routing table updating by exchanging topological information will lead to decreased time for information exchange. Also, application of the store and carry forward approach following the traditional manner shall lead to unnecessary data replication and ultimately data congestion, especially in scenarios of dense traffic. Hence, a compromise needs to be drawn between the value of the exchanged information and the cost that shall be incurred toward its storage.

As a solution to the issues discussed above pertaining to the VDTN implementation, the research on enhancement of such networks indicates an improvement in efficiency of the network through various fixes like optimal relay node placement, employment techniques for the control of nodular congestion and differentiation of traffic according to the capabilities and their network requirements [18].

VDTNs have successfully been implemented in various parts of the world, with a high efficiency rate of data delivery. However, the efficiency of such a network is highly dependent on the participation of vehicular nodes. The cooperation of vehicular nodes toward such data sharing marks the importance that they hold in such a network, often acting as the mobile data carriers for subsequent communication. Such dependency on user cooperation stems from the fact that the routing protocols proposed for such applications are designed with the assumption of full cooperation of the vehicular nodes. The study toward reducing the dependency on the efficiency of VDTN communication solutions on the levels of vehicular cooperation is an area of research that is underway.

Table 2.1 describes the various proposed routing protocols which are employed frequently in VDTNs [18].

TABLE 2.1

VDTN Routing Protocol Summary

Scheme Name	Number of Message Copies	Type	Target
Direct delivery	Single	Direct	Node moves and delivers the packet directly
			Packet is delivered in result of random walk
First contact	Single	Probabilistic	search to its destination
Epidemic routing	Multiple	Blind flooding	Enormous data propagation
			Packets only flooded to the nodes near to the
Surrounding routing	Multiple	Limited flooding	destination
Spray and wait	Multiple	Controlled flooding	Limited copies of packet are generated
			Packets forwarded on the basis of encounter
PRoPHET	Multiple	Probabilistic	History

2.5.4 Delay-Tolerant Networks for Underwater Communications

Underwater activity monitoring is not always possible by personal human intervention. The activities related to underwater operations and systems often need sensors and IoT-based systems. For such a system to work, it needs a connection between these sensory nodes, to form a communication system, with connectivity between the nodes ensuring data transfer to the central data collection node. The DTN provides such a connectivity, taking into consideration its compatibility with mobile nodes and changing topology in network, which might be encountered in underwater monitoring scenarios [28].

Water pollution is a growing concern worldwide that needs to be addressed immediately. As a check on water quality and identifying the various reasons for its degradation, sensor-based networks for such investigation and analysis have been deployed in various water bodies in various countries across the European Union. Instead of an end-to-end-based communication network, realization of such a system is much easier and cheaper through the storage and forwarding scheme of DTN, ultimately using a parent node as data collection center, often housed on a float or a boat that frequents various nodes while cruising over a large area of water, making it an efficient network. Such a network eliminates the requirements of end-to-end connectivity to data centers, as well as lowering the number of data collection centers that might be needed to set up such a network, and hence being cost effective in both deployment and operation, while not sacrificing the efficiency of the application [29].

Marine life monitoring is another important application that DTNs find in the underwater communications scenario. Such an application might serve a large base of studies, ranging from research-based analysis of marine life and the patterns in activity that are followed to keep track of effects that water pollution has on marine life and the exact causes leading to changes in the normal marine patterns as a result of human intervention. An architecture very similar to the one that is implemented in water pollution monitoring is employed here, closely following the topology in that scheme [30].

Deep sea exploration is a costly and dangerous activity. However, that has its own dividends given the environment created at such depths and hence the fauna found there. Human exploration at such depths is often impossible, and may be lethal if attempted. The robotics-based exploration is again a costly affair. It can be made cheaper by employment of exploratory nodes transmitting the data to a sea-level data collector instead of a self-storage capacity laden explorer. Such employments have found their implementations to be productive and cost effective while being efficient in functioning [31].

Some other underwater applications of DTN-based communication and network systems include Navy-based communication and sensory monitoring centers that employ modified forms of DTN networking architecture in their applications.

2.5.5 Military Applications

Military communications systems are one of the most valuable as well as most vital component in any military outfit's array of apparatus. It needs to be robust and safe as well as efficient. The traditional end-to-end network protocols fit the above description to a fair extent and hence can be seen as an option for implementation. However, end-to-end network protocols are not fit for military communication systems, due to the propagation delays, high delivery error rates and the heterogeneous topology that it may employ.

Military communication, along with the above properties, are described with other features, like locations often being remote and needing an on-the-ground communication system due to the security and secrecy concerns toward satellite-based communication. With a decrepit connectivity and remote locations, there is a lot of potential for DTN to be employed as communication systems for military outfits, with adjustments to be made to DTN protocols so as to incorporate the military requirements [32].

Military networks employing DTN have their own challenges. In such a communication network, disruptions in communication and message delivery are very likely and unpredictable in occurrence, given the dynamic nature of topology and situations as well as difficult terrains. There are possible delays due to the radio systems, which have low capacity compared to the massive transmitters and receivers used in common communication systems.

Also, given the inherent delays in communication that characterize the DTN networking scheme, it is not possible to implement such a system as is, due to the urgency and certainty of message delivery being an important aspect in such an application. Hence, for deployment of DTN in such an application, it is amalgamated with important and beneficial features of various technologies, tailor-made for the various situations that such a communication schemes needs to be employed into [33].

DTN has been under study and improvement in the operational and technical design for the military-based applicability solutions, complementing the basic military tactical plan of action and conditioning the network to fit the needs of the battlefield as well as maintain the robustness of the communication system, while maintaining a covert profile for communication technicalities [34]. Many projects have been set up to realize such a goal, so as to promote such research. The foremost in such projects is the Military Disruption Tolerant Networks (MIDNet) project. The MIDNet project's primary aim is to usher in DTN-based solutions with the above-described capabilities. The remodeling of DTN architecture and functionality so as to fit the military requirements is the area of their study. The approach to improvement that guides this project is characterized and parameterized by factors such as robustness in communication factoring in hostile environments, security in message delivery and storage, scalability of network connections on the go and increased data delivery rates, as well as addition of services to the end users.

2.6 Conclusion

Delay-tolerant networks are designed to work in situations where tolerance in delay of information propagation is understandable, trading it for efficiency of data delivery and reliability of network where disruption of network channels, intermittence in information exchange and a general absence of scope or deployment of end-to-end connectivity arise as the major challenges that need to be taken up and addressed.

Delay-tolerant networks have an inherent acceptance of delay and disruption tolerance, handling it with the use of buffer-based implementation of the network where storage and message carry forward methodology is followed, leading to ensuring that data dropping does not occur even in the scenarios of a disrupted network or delay in propagation of said data.

Various routing propositions have been proposed to tackle different challenges that the delay-tolerant networks face. Every routing protocol makes good use of the disruption and delay-prone nature of delay-tolerant networks, enhancing it for the various applications that such networks find themselves in. Such applications of delay-tolerant networks serve a wide range of fields, such as satellite communication using distributed networking, provision and extension of communication to remote areas, allowing and contributing to the growth of intelligent transport system, enhancing terrestrial as well as aquatic environments, and also wildlife monitoring and exploration and military applications, among various other applications.

The delay-tolerant network is an evolving field, where research is being carried out so as to enhance its effectiveness, efficiency and reliability as well as opening newer prospects for its applications. The research to evolve delay-tolerant networks often focuses on challenges like improvement of the data delivery ratio, decrease in delay in propagation, enhancement of routing protocols in terms of intelligence in node-based transfers, addressing buffer overflow and storage issues, taking care of security impediments, making data transfer secure while maintaining the network efficiency and improvement of power efficiency of the network among other concerns.

References

1. Zeng, Deze, Song Guo, and Zixue Cheng. 2011. "The Web of Things: A Survey (Invited Paper)". *Journal of Communications* 6(6): 424–438. doi:10.4304/jcm.6.6.424-438.
2. Ezez. 2019. "Blog: 10 Mind-Boggling Figures that Describe the Internet of Things (IoT)| Cleo". *Cleo*. https://www.cleo.com/blog/internet-of-things-by-the-numbers.

3. Lindgren, Anders, Avri Doria, and Olov Schelén. 2003. "Probabilistic Routing in Intermittently Connected Networks". *ACM SIGMOBILE Mobile Computing and Communications Review* 7(3): 19. doi:10.1145/961268.961272.
4. Mundur, Padma, Sookyoung Lee, and Matthew Seligman. 2009. "Routing in Intermittent Networks Using Storage Domains". *Wireless Communications and Mobile Computing* 11(9): 1213–1225. doi:10.1002/wcm.868.
5. Spyropoulos, T., K. Psounis, and C. S. Raghavendra. 2008. "Efficient Routing in Intermittently Connected Mobile Networks: The Multiple-Copy Case". *IEEE/ACM Transactions on Networking* 16(1): 77–90. doi:10.1109/tnet.2007.897964.
6. Lindgren, A., A. Doria, and O. Schelen. 2003. Probabilistic Routing in Intermittently Connected Networks. In: *Proceedings of the ACM SigMobile*, vol. 7(3), pp. 19–20.
7. Boukerche, Azzedine, and Amir Darehshoorzadeh. 2014. "Opportunistic Routing in Wireless Networks: Models, Algorithms, and Classifications". *ACM Computing Surveys* 47(2): 1–36. doi:10.1145/2635675.
8. Kang, Hyunwoo, and Dongkyun Kim. 2010. "Vector Routing Protocols for Delay Tolerant Networks". *International Journal of Ad Hoc and Ubiquitous Computing* 6(1): 40. doi:10.1504/ijahuc.2010.033824.
9. Yang, Dengfeng, Xueping Li, Rapinder Sawhney, and Xiaorui Wang. 2009. "Geographic and Energy-Aware Routing in Wireless Sensor Networks". *International Journal of Ad Hoc and Ubiquitous Computing* 4(2): 61. doi:10.1504/ijahuc.2009.023897.
10. Fan, Jialu et al. 2013. "Geocommunity-Based Broadcasting for Data Dissemination in Mobile Social Networks". *IEEE Transactions on Parallel and Distributed Systems* 24(4): 734–743. doi:10.1109/tpds.2012.171.
11. Abdelkader, Tamer, K. Naik, A. Nayak, N. Goel, and V. Srivastava. 2013. "SGBR: A Routing Protocol for Delay Tolerant Networks Using Social Grouping". *IEEE Transactions on Parallel and Distributed Systems* 24(12): 2472–2481. doi:10.1109/tpds.2012.235.
12. Cao, Yue, and Zhili Sun. 2013. "Routing in Delay/Disruption Tolerant Networks: A Taxonomy, Survey and Challenges". *IEEE Communications Surveys and Tutorials* 15(2): 654–677. doi:10.1109/surv.2012.042512.00053.
13. Zhu, Ying, Bin Xu, Xinghua Shi, and Yu Wang. 2013. "A Survey of Social-Based Routing in Delay Tolerant Networks: Positive and Negative Social Effects". *IEEE Communications Surveys and Tutorials* 15(1): 387–401. doi:10.1109/surv.2012.032612.00004.
14. Seligman, Matthew, Kevin Fall, and Padma Mundur. 2006. "Alternative Custodians for Congestion Control in Delay Tolerant Networks". In: *Proceedings of the 2006 SIGCOMM Workshop on Challenged Networks—CHANTS '06*. doi:10.1145/1162654.1162660.
15. Silva, Aloizio P., Scott Burleigh, Celso M. Hirata, and Katia Obraczka. 2015. "A Survey on Congestion Control for Delay and Disruption Tolerant Networks". *Ad Hoc Networks* 25: 480–494. doi:10.1016/j.adhoc.2014.07.032.
16. Kate, Aniket, Gregory M. Zaverucha, and Urs Hengartner. 2007. "Anonymity and Security in Delay Tolerant Networks". In: *2007 Third International Conference on Security and Privacy in Communications Networks and the Workshops—Securecomm 2007*. doi:10.1109/seccom.2007.4550373.

17. Farrell, Stephen, and Vinny Cahill. 2019. "Security Considerations in Space and Delay Tolerant Networks". In: *2nd IEEE International Conference on Space Mission Challenges for Information Technology (SMC-IT'06)*. doi:10.1109/smc-it.2006.66.
18. Freimann, A., T. Tzschichholz, M. Schmidt, A. Kleinschrodt, and K. Schilling. 2016. "Applicability of Delay Tolerant Networking to Distributed Satellite Systems". *CEAS Space Journal* 8(4): 323–332. doi:10.1007/s12567-016-0127-3.
19. Caini, Carlo, Haitham Cruickshank, Stephen Farrell, and Mario Marchese. 2011. "Delay- and Disruption-Tolerant Networking (DTN): An Alternative Solution for Future Satellite Networking Applications". *Proceedings of the IEEE* 99(11): 1980–1997. doi:10.1109/jproc.2011.2158378.
20. Ivancic, W., W. M. Eddy, D. Stewart, L. Wood, J. Northam, and C. Jackson. 2010. "Experience with Delay-Tolerant Networking from Orbit". *International Journal of Satellite Communications and Networking* 28(5–6): 335–351. doi:10.1002/sat.966.
21. Burleigh, S., A. Hooke, L. Torgerson, K. Fall, V. Cerf, B. Durst, K. Scott, and H. Weiss. 2003. "Delay-Tolerant Networking: An Approach to Interplanetary Internet". *IEEE Communications Magazine* 41(6): 128–136. doi:10.1109/mcom.2003.1204759.
22. Murillo, Martin J., and Mozafar Aukin. 2011. "Application of Wireless Sensor Nodes to a Delay-Tolerant Health and Environmental Data Communication System in Remote Communities". In: *2011 IEEE Global Humanitarian Technology Conference*. doi:10.1109/ghtc.2011.97.
23. Ntareme, Hervé, Marco Zennaro, and Björn Pehrson. 2011. "Delay Tolerant Network on Smartphones". In: *Proceedings of the 3rd Extreme Conference on Communication the Amazon Expedition—Extremecom '11*. doi:10.1145/2414393.2414407.
24. Fall, Kevin. 2003. "A Delay-Tolerant Network Architecture for Challenged Internets". In: *Proceedings of the 2003 Conference on Applications, Technologies, Architectures, and Protocols for Computer Communications—SIGCOMM '03*. doi:10.1145/863955.863960.
25. Isento, Joao N. G., Joel J. P. C. Rodrigues, Joao A. F. F. Dias, Maicke C. G. Paula, and Alexey Vinel. 2013. "Vehicular Delay-Tolerant Networks? A Novel Solution for Vehicular Communications. *IEEE Intelligent Transportation Systems Magazine* 5(4): 10–19. doi:10.1109/mits.2013.2267625.
26. Soares, Vasco N. G. J., Farid Farahmand, and Joel J. P. C. Rodrigues. 2009. "A Layered Architecture for Vehicular Delay-Tolerant Networks". In: *2009 IEEE Symposium on Computers and Communications*. doi:10.1109/iscc.2009.5202332.
27. Bhardwaj, Kartik Krishna, Anirudh Khanna, Deepak Kumar Sharma, and Anshuman Chhabra. 2019. "Designing Energy-Efficient Iot-Based Intelligent Transport System: Need, Architecture, Characteristics, Challenges, and Applications". *Energy Conservation for Iot Devices* 206: 209–233. doi:10.1007/978-981-13-7399-2_9.
28. Zeng, Yuanyuan, Kai Xiang, Deshi Li, and Athanasios V. Vasilakos. 2012. "Directional Routing and Scheduling for Green Vehicular Delay Tolerant Networks". *Wireless Networks* 19(2): 161–173. doi:10.1007/s11276-012-0457-9.
29. Chan, Colin Y. M., and Mehul Motani. 2007. "An Integrated Energy Efficient Data Retrieval Protocol for Underwater Delay Tolerant Networks". *OCEANS 2007 - Europe*. doi:10.1109/oceanse.2007.4302341.

30. Balas, Valentina E., Le Hoang Son, Sudan Jha, Manju Khari, and Raghvendra Kumar. n.d. *Internet of Things in Biomedical Engineering*.
31. Guo, Xiaoxing, Michael R. Frater, and Michael J. Ryan. 2006. "A Propagation-Delay-Tolerant Collision Avoidance Protocol for Underwater Acoustic Sensor Networks". *OCEANS 2006—Asia Pacific*. doi:10.1109/oceansap.2006. 4393849.
32. Partan, Jim, Jim Kurose, and Brian Neil Levine. 2007. "A Survey of Practical Issues in Underwater Networks". *ACM SIGMOBILE Mobile Computing and Communications Review* 11(4): 23. doi:10.1145/1347364.1347372.
33. Lu, Ziyi, and Jianhua Fan. 2010. "Delay/Disruption Tolerant Network and Its Application in Military Communications". In: *2010 International Conference on Computer Design and Applications*. doi:10.1109/iccda.2010.5541302.
34. Bekmezci, Ilker, and Fatih Alagöz. 2009. "Energy Efficient, Delay Sensitive, Fault Tolerant Wireless Sensor Network for Military Monitoring". *International Journal of Distributed Sensor Networks* 5(6): 729–747. doi:10.1080/15501320902768625.

3

Deep Learning-Based Decision-Making with WoT for Smart City Development

S. Vimal, V. Jeyabalaraja, P. Subbulakshmi, A. Suresh,
M. Kaliappan and S. Koteeswaran

CONTENTS

3.1 Introduction

In the real world, there have been a lot of significant transformations from the modern digital world, including technology-driven development and its implementation in public valued spaces. Threats happening in the technocracy have a major impact in the public space providing a realization of the short growth in the values of the complex quality of life. The environment in which the technological developments take place is a complex network with the actors operating as islands to illustrate the smart city. The networked environment has a smart city ecosystem that requires stakeholders with different mindsets to be analyzed in the process of knowledge transfer, and knowledge from the proper decision-making has also been analyzed. The decision-making has the smart city ecosystem with its technology focusing on value adding and conflict of values, as shown in Figure 3.1. The decision-making is assumed to be faulty in nature, based on the technological factors in the situations that create the assessments with the technological society and it has been held to be the balancing mechanism helping to assess the renewed implementation.

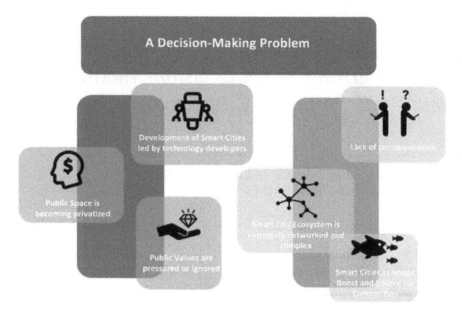

FIGURE 3.1
Decision-making for urban planning.

The IoT-based industry has enormous growth in urban planning with the foundation in development of industry for chips, microcontroller kits, electronic gadgets and telecommunication systems forming a part of the industrial segments. Attacks and vulnerability issues are the main concern in the IoT with the network breach. The vulnerability can be inspected with Intrusion Detection Systems that can encounter the anomalies in the reliable services of IoT applications. Various deep learning models have been proposed, with the modeling scheme of learning and data from sensors forming the structure of the setup. The internet is used for sharing of communication and content with a global communication establishing the objects connected and the smart devices with the support of international data corporations (IDCs) [1]. IoT paved the way for smart cities with support in optimization and enhancement of the public services in a smarter way, leading to smart transportation, smart parking, smart hospitals and urban development. The city offers a strategy of control over various domains established with a new service rule for the automatic classification of the services offered. Data from different sources are being collected with the particular location and time [2]. The expansion of cities handles the smarter way with smarter development in a cloud-based environment.

In an edge computing-based cloud setup, IoT data gathering includes all data to be gathered from the remote server. Remote processing has been done in an emerging area with bandwidth usage with an increased speed. Fog computing and data processing have been done from the remote servers that increase the benefits of the data size and the low latency simulation in the filtered server data. Edge computing offers very large offloading and the storage accuracy has been enriched with privacy preservation in a smart way using IoT prediction [3].

In the modern era, people are getting connected to the real world through the online sector, having an integrated circuit technology and wireless communication that have been established using signal acquisition and data preprocessing with the wireless communication capabilities [22]. The IoT has been divided into three layers: the perception layer, network layer and application layer [4]. The perception layer deals with the end devices and tail nodes. The end devices monitor a progressive approach in elaborating the segment with the layer channel in the devices to support the RFID tags, cameras and GPS devices. The sensors are used by the parameters to form an environment in the sensing layer with the information from the physical world with analog-to-digital conversion [5]. The second phase of the IoT deals with the network layer, which connects the sensing layer with the application layer, and which includes mobile communication and fixed communication using internet and private communication networks. The network layer includes various functionalities includes optical fiber communication, security access and satellite access [6]. The network layer deals with the applications in application layer; human–computer interaction (HCI) is one of the biggest applications that retrieves the data from the valuable environment. The end devices include the IoT data retrieval from the environment with an information security and business domain. The security includes the analysis of data, computation, storage and various technologies. The transmission of data communication is done between the end devices and the sensors [7].

IoT brings the solution for the problem prevalent in the recent stochastic applications to the problems on recent surveys on the existing RFID technology. In the next five years, it is assumed the usage of IoT will cross the range of 30 billion [8]. RFID will reach more than a billion in the next few years due to drastic changes in the development of IoT devices like sensors, antennas and circuit development using semiconductors. The security of IoT includes authentication with the secure hash key toward intrusion detection and external data behavior with Intrusion Detection Systems. IoT security enables the system to enhance hotspots in the system. Figure 3.1 shows the layered structure of IoT in a Wi-Fi management system [9].

The perspectives of IoT enhance the development of IoT in all levels of the industry. A survey indicates that China developed an economy in the infrastructure to expand Chinese industry [10]. There are a limited number of uses of the applications in the market for the government, with private investment being the real market under investigation [28]. The rise of IoT is hugely involved in smart homes with the WLANs dealing with the public market. Intrusion detection involves faster detection with the cost increasing with the benefits. Intrusion detection has a major involvement in wireless communication technology and the sensing assumes to be of a sensing assumption. Various long-distance communications have been established with the various remote sensing applications in conjunction with the sensors in UART, LoRa, ZigBee and Bluetooth applications [11]. Ubiquitous Wi-Fi includes a widespread application of IoT security-based technology and also the security limitations of monitoring the usage of security with the hotspot in the current IoT [12]. The economic deployment has the option of considering the data security with the deep learning algorithms achieving a remarkable role in IoT considerations. The deep learning involves data sampling of the content in the real world. Deep learning is achieved from the problem, with the target data having been assumed in the development of the deep neural networks [13]. The Chinese IoT has its own development in the IoT, with various research universities vying to be the university to communicate the performance in the classification of the RFID R&D [14]. The image classification depends on the industry in the Federation of China which has a migration algorithm with the parameter to be analyzed. The image segments are being nowadays used in analyzing cancer predictions for smart city establishment in recent paradigms, using the IoT wearable devices and classifiers [29]. The layered architecture is represented in Figure 3.2.

The security technology has an authentication technology to provide a secure transmission with the passive defenses, and the state, behavior and usage may be provided with the IoT security in devices [15]. The intrusion uses a parameter to detect the unauthorized behavior. The function of IoT has a wireless LAN which has a convenient environment that can automatically provide the humidity and adjust the room temperature, switching due to the context awareness [16]. The wireless communication technology has a big opportunity in the simultaneous sensing of the Wi-Fi signals in the same paradigm [25]. The information sensed can be

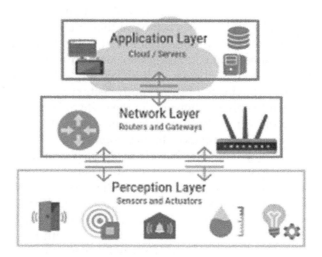

FIGURE 3.2
Layered services for WoT.

assumed to be in the additional communication that can be done in the Wi-Fi establishment [17].

The learning categories that became the essential methods of treating data are inductive learning, direct learning and supervised learning [18,26]. Inductive learning has a high detection rate with abnormal data. The dimensions of the training data and the test data have to be consumed to fit the data. The inverse of the homogeneity problems, regression of the data can be directly assessed with the existing research. Direct push learning is one of the techniques in supervised learning and is effectively addressed in the tasks of the unauthorized learning's clear characteristics. Intrusion is an unauthorized operation of accessing, destroying and tampering with the causes and damages prevailing in the system [19]. The safety of resources and the sensors are in addition to the physical sensor protection. The first line of defense uses data encryption and the sensor network security; key management and the protocols have been identified for the authorized parties [20].

The IoT sensor has attacker nodes, and the weak points of the system have IoT sensors with defensive nodes [21]. The IoT has an intensified proposal for the active defense system with the attacker being established in the automatic mechanism in the sensor network. The active defense nature has the attacker's behavior and the intrusion information in the most secure environment [27]. The IDS has the best feature in the detection in the source/sink with the nodes having a secure environment. It uses an automation system with the attack behavior, which has low traffic in the short wireless communication and an alarm signal with the node-based distributed architecture.

The distributed environment of the IoT, IDS has the collaborative condition in the IoT, with the specific attacks in the predetermined area IDS. The IoT helps to discover various attacks within the sensor nodes with the generation of agents in a communication activity in a specific location [23,24]. IDS agents are pre-located within the neighboring node in the wireless communication channels, and are found in interactive models with separated data between areas, on the basis of better learning in a complete target field for the service of the specific locations. The learning tasks are allocated with greater expression within the target field.

3.2 WoT-Based Decision-Making for Smart City Development

The smart ecosystem has a phase of contact in the value ecosystem to relate to the biological factor that has been used in the animal survey of the complex relationship. The biological relationships include the stakeholders with high/low positions and choice-based networks in cities with dynamic events that have been planned in the complex network structure [15]. The responsible innovation has been utilized by various researchers and academics, and the public values are focused on the stress factors in the responsible innovation and on the transparent manner of the actors' phase, with a view of the products in the acceptability and sustainability of societal desirability in the scientific and technological factors of the marketable products. It follows three dimensions in innovation with the following factors:

1. Dimension of anticipation: The way has been incorporated in the innovation with the benefits and the impacts.
2. Reflexive action: The stakeholders' intentions, interests, values and the actors involved are prescribed with the WoT-based smart city's progress. This dimension shows the innovation in the adaptable societal needs to be fulfilled with the interest planned in the great challenges in the environment.
3. Better response: The dimension with this innovation shows the responsiveness to societal needs, with the interest of forecasting the grand challenges with the new ideas and the innovation of creation.

The better response paves the way for decision-making in an important role to focus on the dimensions in both an active and a passive way. The responsibility can be viewed with the way of decision-making in the evaluation pro forma. The passive mode of responsibility is assumed to be the structures leading to a desirable outcome, in the event of an undesirable mode following the focus on the intense way of treating the punishment [18]. The public organization works on the active way of ensuring responsible outcomes of the decision-making in the municipal organizations and the active responsible organizations.

Technology assessment is done in alignment with the technology to carry out a constructive assessment of the tools used to design the new technologies in the decision-makers' process. Following the technology assessment in the new design practice, with the users to be predicted from the start and interactive communities to explore new trends in societal learning, the technology is assessed to be the higher of the chance factors in the significant adjustment to the responsible innovation way of enabling the technology, in the order of processing the innovation in the singular approach to the designated approach in the formulated condition.

3.3 Value Conflicts and Analysis of Constructive Technology Assessment (CTA)

The sociotechnical value map is a tool that assesses the performance of the research tool with the technology assessment of the value sensitive design.

Public value has been assumed to be that of the specific users of the technology in the public standard. The technology assessment in public value has been assumed to be the public construction in CTA (Constructive Technology Assessment) frameworks in a specific set of actors in the VSD (Value Sensitive Design) practices in the selected and right combination for the value-based approach in the design process. The smart city has been exposed to the sociotechnical public with the representation in the recommendation in the policy makers to incorporate values in the design for the smart cities with the continuous technology assessment. Researchers are used to mapping identical values and situations with the use of technology problems that shift over time. It is not a static system, but it is a dynamic mapping of the smart ecosystem in a rational way, analyzing the possible future scenarios in smart city management. The smart ecosystem has various factors and temporal characteristics, as shown in Figure 3.3. The demonstrations, presentations and analysis have been discussed with the stakeholders.

3.4 Decision-Planning Using WoT in Urban Ecosystems with Deep Learning

Urban planners and decision-makers are facing a huge flow of information prevailing in the data center. The smart city focus has been imposed upon the urban ecosystem to set up WoT-based system settings in the decision policy. The information access has enormous assumptions in the period of evolving the system in the data production and dissemination of field policy, and the human access to the information is a great challenge in the repository in the bits arrival. The urban planners and decision-makers face a rapid and enormous growth that has been provided in the urban environment. Changes to the urban system may come from a sudden error following the urban planning decisions, and understanding these decisions' consequences increases the information and planning needed. The desired future means the decision-makers have been awareness planners in various parts of society. Urban planning requires a more holistic understanding and awareness of the situation. The urban elements have the subsystems with their complex influence on each other and the situational awareness becomes increasingly necessary to maintain the competence of single individual coordinating the dynamic systems. The urban area has been challenged in the interactive method with the participation of small city planning in the city representation of the imagined phenomenon. Digitalization and participation are the main platforms for smart city processing with scientific reasoning. The decision-making role has been well established in the communicative theory processing with the roles of various stakeholders in the participative and collaborative process in the theory of a dignified process. The acknowledgment

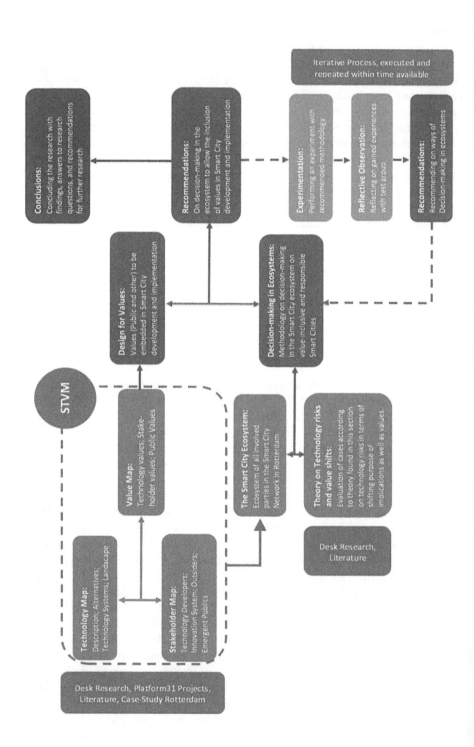

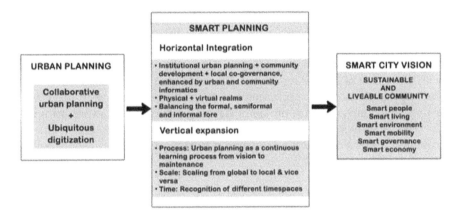

FIGURE 3.4
Collaborative approach for a smart city.

must be made for the stakeholders and participants in the planning process as early on possible. Digitalization has been done in urban planning and decision-making to adopt the new transdisciplinary route for urban decision-making practice. The smart city uses ICT to optimize the efficiency of the useful, necessary city processes. The smart cities have the collective intelligence with the situational awareness in the city. The smart city phenomenon has been severely criticized, for instance, for ready optimizations between the good and bad cities. The problem with the smart city option is that it is meant for the urban planning in the holistic and integrative perspectives and the multiscalar stress planning is done in the forum as shown in Figure 3.4. The urban planning approach is done on behalf of the planning theories and the post-structural theories that comprise the monitoring and the evaluation in the planning process. The urban planning has been created with the purpose of scaling the horizontal and vertical participation.

3.5 RESTful Services Using WoT

A web API is a development in web services where the emphasis is on moving to simpler representational state transfer (REST)-based communications. RESTful APIs do not require XML-based web service protocols (SOAP and WSDL) to support their interfaces. RESTful web services offer an interoperable and secure connection to maintain the data in the M2M (machine-to-machine) connections and communications on the internet based on the classes done on the representational state, and transfer within the compliant web services where the resources have been manipulated with the uniform set of state and stateless transfers emerging within the web services' control in the arbitrary set of events, and the option of the web usage has been utilized in the context of IoT within the arbitrary web services. The resources are assumed to be of the representational condition in WoT within the deep learning context, and the web services have a multivariate option of analyzing the maximum prediction in the resolved unit. The http-based modern application has been used with the web services to analyze the context within the deep learning context of modular thinking. The web services using the SOAP (Simple Object Access Protocol) platform provide a modern context for the urban planning data center to contribute to the smart city essentials.

3.6 Conclusion

The complex urban planning ecosystem measures the overall development of the process to construct a clear, defined area for the challenging task of producing the planners and the decision-making with the situational awareness of the planned ecosystem. Deep learning-based decision-making with WoT for smart city development has started the phase with its implementation in urban planning, and this task making may help the future planners to construct an awareness of the process and research changes to the own fittings that allow the problem to meet the needs of the user with the formation of the knowledge management process in the information context. The digitalization and urbanization fulfill the objective of policy-makers in the integration of the collaborative methods. The system thinking involves the perspectives of learning the situational awareness of complex ecosystems and brings the data together to form a knowledge-based deep learning supportive smart city planning with a collaborative approach. The stakeholders involved in the development process bring diverse thinking to the process of getting the channels with a human-centric focus on planning.

References

1. Alvi S, Shah G, Mahmood W (2015) Energy efficient green routing protocol for Internet of multimedia things. In: *Proceedings of the IEEE 10th International Conference on Intelligent Sensors, Sensor Networks and Information Processing*, Singapore.
2. An J, Gui X, Yang J, Sun Y, He X (2015) Mobile crowd sensing for Internet of Things: A credible crowdsourcing model in mobile-sense service. In: *Proceedings of the IEEE International Conference on Multimedia Big Data*, Beijing, China, p 92–99.

3. Andersson M (2015) Short range low power wireless devices and Internet of Things (IoT), White Paper UBX-14054570. *U-blox.*
4. Khan Z, Anjum A, Kiani SL (2013) Cloud based big data analytics for smart future cities. In: *Proceedings of the IEEE/ACM 6th International Conference on Utility and Cloud Computing.* IEEE Computer Society, p 381–386.
5. Fan W, Bifet A (2013) Mining big data: Current status, and forecast to the future. *ACM SIGKDD Explorations Newsletter* 14(2):1–5.
6. Bertot JC, Choi H (2013) Big data and e-government: Issues, policies, and recommendations. In: *Proceedings of the 14th Annual International Conference on Digital Government Research.* ACM, p 1–10.
7. Amin R, Kumar N, Biswas G, Iqbal R, Chang V (2018) A light weight authentication protocol for IoT-enabled devices in distributed cloud computing environment. *Future Generation Computing Systems* 78:1005–1019.
8. Neirotti P, De Marco A, Cagliano AC, Mangano G, Scorrano F (2014) Current trends in Smart City initiatives: Some stylised facts. *Cities* 38:25–36.
9. Yin J, Sharma P, Gorton I, Akyoli B (2013) Large-scale data challenges in future power grids. In: *2013 IEEE 7th International Symposium on Service Oriented System Engineering (SOSE).* IEEE, p 324–328.
10. Eräranta S (2013) Situation awareness in urban planning. Case: Mobility planning decision-making in Otaniemi Campus. Master's thesis. Aalto University, Espoo Finland.
11. Griffinger R, Fertner C, Karmar H, et al. (2007) Smart cities: Ranking of European medium-sized cities. Available at http://www.smartcities.eu/down load/smart_cities_final_report.pdf., February 2015.
12. Silva CN (2010) The E-planning paradigm – Theory, methods and tools: An overview. In: Silva CN (ed) *Handbook of Research on E-Planning: ICTs for Urban Development and Monitoring.* IGI Global, Hershey, PA, p 1–14.
13. Staffans A, Horelli L (2014) Expanded urban planning as a vehicle for understanding and shaping smart, liveable cities. *The Journal of Community Informatics* 10(3).
14. van Hoek M, Wigmans G (2011) Management of urban development. In: Franzen A, Hobma F, de Jonge H, et al. (eds) *Management of Urban Development Processes in the Netherlands – Governance, Design, Feasibility.* Techne Press, Amsterdam, p 53–76.
15. Vanolo A (2014) Smartmentality: The smart city as disciplinary strategy. *Urban Studies* 51(5):883–898.
16. van't Verlaat J, Wigmans G (2011) Introduction. In: Franzen A, Hobma F, de Jonge H, et al. (eds) *Management of Urban Development Processes in the Netherlands – Governance, Design, Feasibility.* Techne Press, Amsterdam, p 17–32.
17. Viitanen J, Kingston R (2014) Smart cities and green growth: Outsourcing democratic and environmental resilience to the global technology sector. *Environment and Planning Part A* 46(4):803–819.
18. Väyrynen E (2010) *Towards an Innovative Process of Networked Development for a New Urban Area. Four Theoretical Approaches.* Studies in Architecture 2010/43, Aalto University, Espoo.
19. Te Brömmelstroet M, Bertolini L (2008) Developing land use and transport PSS: Meaningful information through a dialogue between modelers and planners. *Transport Policy* 15(4):251–259.
20. Townsend AM (2013) *Smart Cities: Big Data. Civic Hackers, and the Quest for a New Utopia.* W.W. Norton Company, New York.

21. Silva CN (2010) The E-planning paradigm – Theory, methods and tools: An overview. In: Silva CN (ed) *Handbook of Research on E-Planning: ICTs for Urban Development and Monitoring*. IGI Global, Hershey, PA, p 1–14.
22. Hong Liu, Ning H, Mu Q, et al. (2019) A review of the smart world. *Future Generation Computer Systems* 96:678–691.
23. Vimal S, Kalaivani L, Kaliappan M.,Suresh A., Xiao-Zhi Gao Varatharajan R (2018) Development of secured data transmission using machine learning based discrete time partial observed Markov model and energy optimization in cognitive radio networks. *Neural Computing and Applications*. doi:10.1007/s00521-018-3788-3.
24. Vimal S, Kalaivani L, Kaliappan M (2017) Collaborative approach on mitigating spectrum sensing data hijack attack and dynamic spectrum allocation based on CASG modeling in wireless cognitive radio networks. *Cluster Computing*. doi:10.1007/s10586-017-1092-0.
25. Vimal S, et al. (2016) Secure data packet transmission in MANET using enhanced identity-based cryptography. *International Journal of New Technologies in Science and Engineering* 3(12):35–42.
26. Mariappan E, Kaliappan M, Vimal S (2016) Energy efficient routing protocol using Grover's searching algorithm using MANET. *Asian Journal of Information Technology* 15(24): 4986–4994.
27. Ilango SS, Vimal S, Kaliappan M, et al. (2018) Optimization using artificial bee colony based clustering approach for big data. *Cluster Computing*. doi:10.1007/s10586-017-1571-3.
28. Kannan N, Sivasubramanian S, Kaliappan M, Vimal S, Suresh A (2018) Predictive big data analytic on demonetization data using support vector machine. *Cluster Computing*. doi:10.1007/s10586-018-2384-8.
29. Geetha R, Sivasubramanian S, Kaliappan M, Vimal S, Annamalai S (2019) Cervical cancer identification with synthetic minority oversampling technique and PCA analysis using random forest classifier. *Journal of Medical Systems* 43(9):286. doi:10.1007/s10916-019-1402-6.

4

Predicting Epilepsy Seizures Using Machine Learning and IoT

Bahubali Shiragapur, Tanuja S. Dhope (Shendkar),
Dina Simunic, Vijayalaxmi Jain and Nishikant Surwade

CONTENTS

4.1 Introduction

As per the World Health Organization (WHO) survey, around 60 million people are affected by epilepsy diseases [1]. This is a brain disorder which needs to predict occurrences to prevent life-threatening situations. Epileptic seizures are a result of sudden changes in electroencephalogram (EEG) signals reflected as transient high peaks in EEGs [2]. The EEG signal is widely used for clinical assessments for measuring brain activity and detection of seizures. Traditional, visual scanning of a patient's EEG data is a tedious and time-consuming process. Thus, there is a need for a reliable and automated system to predict, classify and detect epilepsy, for better treatment. Furthermore, this kind of system can reduce the effects of long-term treatment with antiepileptic drugs which are harmful to the human neurological system. Thus, there is a need identify a better predictive system that will help to doctors to make decisions and will reduce clinical observation errors.

A seizure represents a single occurrence event in EEG. However, epilepsy is defined as a neurological state characterized by two or more unprovoked seizures. Apart from various types of seizures, most common seizures can be classified as generalized, or in the form known as focal [3,4]. During focal (or partial) seizures, the seizure activity is restricted to a portion of one brain hemisphere. The seizure begins in a part of the brain. Focal seizures are of two types: A simple partial seizure is a focal seizure with retained awareness. A complex partial seizure or focal dyscognitive seizure is a seizure with loss of awareness.

The EEG is a biomedical clinical tool used to predict human brain abnormalities. Furthermore, the use of better state-of-the-art techniques, such as the 10–20 international system, are used to record brain activity [11,12]. This multichannel time variant signal can be further analyzed by using a signal processing tool to extract different signal features. It is possible to detect and predict epilepsy. This study's primary goal is to detect and predict by the use of a machine learning approach and to process the data by Internet of Things (IoT) devices. Furthermore, this chapter may motivate the research group to solve the societal problem.

We have processed the publicly available EEG signals of normal as well as epileptic disorder patients using Chebyshev filter wavelet analysis, and extracted the features using wavelet decomposition that captured the frequency of the dataset. The seven different techniques, including LPC (linear predictive coding), kurtosis, mean, auto-correlation, skewness, spectral energy and feature extraction is done by using PCA (Principal Component Analysis). The state-of-the-art methods such as the Artificial Neural Network (ANN) are used to make decisions on medical events whether the EEG signal is free from epileptic seizure or not.

4.2 Literature Survey

A set of several unique features can be used for predicting the preictal state of epileptic seizures. Rasekhi et al. [5] have proposed univariate linear feature detection for seizure prediction; 132-dimensional feature space has been used by utilizing six EEG channels and extracting various univariate linear properties. The work suggests that the "preictal time" begins 10 to 40 minutes ahead of the "ictal" state, with a difference of 10 minutes. "Preictal" and "ictal" states are considered as a binary classification tests to predict an epileptic disorder. They reported that prediction sensitivity is 73.9%. They used a Support Vector Machine (SVM) for the classification of EEG signals "preictal" and "ictal." The authors suggested the use of univariate linear features with a fixed window size, with certain regularization decisions on EEG signals. They have considered 5-second segmented data for further processing and filtering to reduce noise.

Teixeira et al. [6] have developed a technique for predicting the epileptic seizure in real time. The medical event prediction is based on machine learning classification methods, viz. SVM, radial basis function (RBF) neural networks and multilayer perception (MLPs) neural networks. The authors [7] have suggested filtering techniques as a preprocessing step, which can be used for removal of artifacts. They suggested the SVM classifier can achieve seizure detection sensitivity of 75.8% [8]. The use of the wavelet signal processing method for prediction of seizures has been suggested in Gadhoumi et al. [9]. The wavelet energy and wavelet entropy are considered as features to train the neural network. For testing, six patients' data was selected from two or three channels. The reported sensitivity is 88%.

Zandi et al. [10] have also suggested a model for prediction of seizures based on zero crossing by scalp EEG signals. The computation of histogram for all intervals in moving the average window for observations at different sets of points (preictal and interictal). The work suggests an alarm is created at the start or beginning of the "preictal" state of the seizure, which predicts seizures. The suggested model provides a sensitivity of 88.3%, with a predicted time of 22.5 minutes.

The authors [12] suggested that statistical moments, the first four as features, are extracted. These features are used to measure the variance, similarity and symmetry of successive EEG signal samples. As the EEG is a non-stationary signal, it is necessary to eliminate noise smoothing of EEG signal for better sensitivity and performance analysis of epilepsy EEG dataset signals.

The authors [13] have shown the experiment for detection of epileptic seizure by using filtering and wavelet transformation techniques. They [14] have presented a classification of sleep stages operating on wavelet transformation and neural networks. The detailed and db-4 approximation coefficients and back-propagation algorithm EEG are used to train the neural network model for classifying the stages of sleep signals [18]. The authors [16] have proposed the use of lower devices MSP-432 for seizure detection. Another research group [17] discussed accurate detection of epilepsy with a 10-second prediction time well before the occurrence of the medical event.

4.3 Proposed Approach

Figure 4.1 shows the block diagram of the proposed approach.

4.3.1 EEG Dataset

A public dataset [15] containing five EEG sets are used for experimentation. The details of the dataset are as follows:

- Signal recording segments: 100
- Sampling frequency: 1,000 Hz
- Duration: 23.6 seconds of every segment

For experimentation we have used only two sets: A, E.

Healthy patients' EEG signals are in Set-A, whereas epileptic patients' EEG signals during epilepsy seizures are in Set-E.

4.3.2 Preprocessing

The Chebyshev filtering method is used to remove the artifacts of EEG signal as it is a non-stationary signal. This will enhance the prediction and detection of the epilepsy seizure event.

4.3.3 Decomposition

Decomposition is a sub-band coding that uses Daubechies wavelet families to decompose the signal for analysis of low frequency band signal. For experimentation, the db-4 method is used for analysis of the EEG dataset.

4.3.4 Feature Extraction

Various statistical methods are considered, such as LPC, kurtosis, mean, auto-correlation and PCA. These features are considered to train the network for further classification.

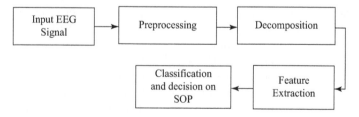

FIGURE 4.1
Block diagram of the proposed system.

4.4 Linear Predictive Coding (LPC)

The LPC coefficients were computed for the non-stationary EEG dataset. Thirteen LPC coefficients are considered for simulation. In linear prediction, the present EEG samples are approximated to the linear combination of past samples, that is:

$$s(n) = \sum_{k=1}^{p} \alpha_k s(n-k) \text{ for some given value of } \rho, S, \alpha_\kappa \qquad (4.1)$$

In the above equation:
α_κ, s: Linear prediction coefficients
$s(n)$: Windowed speech sequence

$s(n)$ is calculated by the product of a short time speech frame with either hamming or a similar type of window:

$$S(n) = x(n){}^*w(n) \qquad (4.2)$$

In the above, the windowing sequence is represented by $\omega(n)$.

4.5 Kurtosis

Kurtosis is used to find heavy-tiredness or light-tiredness of data when compared to a normal distribution. High kurtosis indicates heavy tails and similarly with low kurtosis indicates light tails. For data $Y_1, Y_2 \ldots Y_N$, the kurtosis is given by:

$$\text{Kurtosis} = \frac{\sum_{i=1}^{N} (Y_i - Y)^4 / N}{S^4} \qquad (4.3)$$

In the above equation:
S represents standard deviation
Y represents mean
N represents the number of data points

To calculate standard deviation, N is used for the denominator in kurtosis.

4.6 Mean

The average of the EEG signal is used as another statistical parameter for multichannel dataset signals. This feature helps to differentiate epileptic seizure event.

4.7 Classification

Classification deals with classifying a new observation; it is based on knowledge gained from training data with known category membership. Various methods, such as neural networks like back propagation, LVQ, SOM, feed forward, normalized correlation, K-nearest neighbor (KNN), Hamming distance, support vector machine (SVM), weighted Euclidean distance, which are used for classification.

In this chapter, SVM and KNN are used for pattern classification, where the observations are classified based on features.

4.8 K-Nearest Neighbor (KNN)

KNN is one of the popular classification techniques because of its simplicity and robustness. The test observation (feature) is classified by finding a nearest neighbor calculated from training observations (features).

Various distance metric measurement techniques such as Euclidean distance, Chebyshev, city block, correlation, cosine, Spearman, Hamming, and so on, are used to find the distance between the training and testing vectors.

A simple Euclidean distance formula for the distance between training and testing vectors is given by:

$$d(a,b) = \sqrt{\sum_{i=1}^{n} (a_i - b_i)^2} \qquad (4.4)$$

The testing vector is assigned with the label of the class which is at the smallest distance.

The features are extracted for training and testing set of observations, representing these observations in some different dimension space. The similarity of two different points can be represented by the distance between them in a space.

The working of the K-nearest neighbor algorithm is as follows [11–13]:

- Define a positive integer value K. Also calculate the features of the new observation.
- For the new observation, calculate the K closest distances.
- Find the closer classification of this training observation.
- This helps us to classify the new observation.

If the result is not satisfactory, change the value of K until the reasonable level of correctness is achieved.

In Figure 4.2, there are two classes, represented by squares and triangles. The new testing feature is introduced in the feature set. The testing feature is classified into the right class using K-nearest neighbor by assigning the label of the higher majority neighbors.

$$A := \frac{1}{n} \sum_{i=1}^{n} a_i \tag{4.5}$$

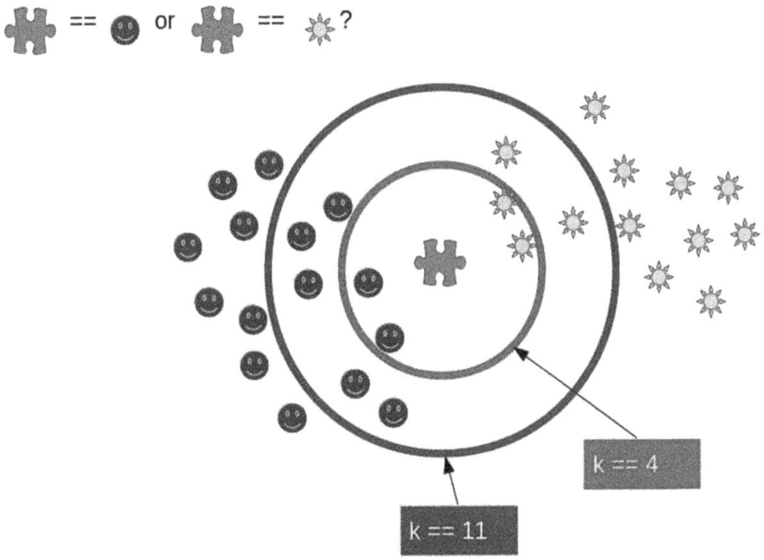

FIGURE 4.2
Classification of query image through a KNN classifier [19].

4.9 Auto-Correlation

The self-correlation of signal is an important feature used to correlate signals based on time difference and average value at origin.

$$R_{XX} = \sum_{i=0}^{\infty} S(n) \qquad (4.6)$$

The correlation coefficient R is used to indicate a higher degree of similarity, [−1, 1], where perfect correlation is given by 1 and perfect anti-correlation is given by −1.

4.10 Principal Component Analysis (PCA)

For low artifact signal analysis PCA is mostly used compared to linear discriminant analysis (LDA) and independent component analysis (ICA). Furthermore, PCA is used for mapping data to lower-dimensional space from high-dimensional space.

4.11 Seizure Prediction

Figure 4.3 shows various medical event timelines, including the "no seizure occurrence" time interval referred to as "interictal."

For preictal, the alarm will be set in forthcoming elapsed time denoted by the Seizure Prediction Horizon (SPH). The Seizure Occurrence Period (SOP) follows the SPH. The seizure is expected to occur at SOP. After the seizure period, during the postictal period, no seizures are observed.

The thumb rule to avoid a harmful situation is that the cumulative time interval of SPH and SOP should be longer than 5 minutes.

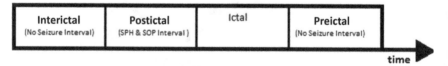

FIGURE 4.3
Seizure event time activity.

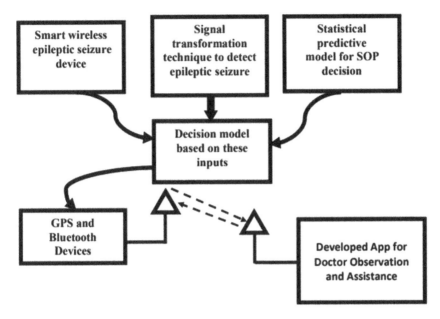

FIGURE 4.4
Web of Things (WoT) to predict epileptic seizure.

The main objective is to identify and detect SOP on the patient side by wearable devices. Further, this data is processed remotely with the help of IoT. The block diagram shows complete wireless Smart Web of Things (SWOT) architecture to monitor epileptic seizure activity at the user end with a comfortable and portable prototype. The end-user observatory device will be connected with the clinical side through existing smartphone devices. Figure 4.4 shows the block diagram of epileptic seizure prediction and detection using signal transform and statistical methods.

Furthermore, such kinds of smart medical devices are today's need, wherein we can monitor health status remotely and also the location of the patient can be traced. The alarm events generated can be further shared to family members as well as to clinical observers for preventive action.

4.12 Results and Discussion

For the simulation analysis, we used the MATLAB tool and the parameters in Table 4.1 are considered. In the proposed work, the EEG signals (Healthy person dataset Set-A and epileptic seizure person dataset Set-E) are pre-processed through the Chebyshev filter. The filtered signal is then decomposed by the Discrete Wavelet Transform. Different statistical features were extracted to differentiate the normal and epileptic EEG signal, as shown in Figures 4.5 and 4.6. The outcomes of the proposed system are described below in a qualitative and quantitative manner.

In the qualitative analysis, the graphical results of the EEG signals are presented.

Figure 4.7 shows the seizure timeline activity signal; we need to look at the adaptive seizure prediction algorithm (ASPA) for better real-time performance.

In Table 4.2, the extracted features using five different techniques have been tabulated. Patients 1–4 are the normal patients while Patients 4–6 are the epileptic patients. From Table 4.2, it is observed that the values of the features of the epileptic and normal patients differ. These features are then applied to the Machine Learning Algorithm (KNN). We have programed and designed the Graphical User Interface (GUI) to display the final output whether the person has epilepsy or not. The results are shown in Figures 4.8 to 4.11.

TABLE 4.1

Simulation Parameter

Descriptions	Parameters
EEG dataset	Set-A and Set-E [15]
Filter	Chebyshev filter
Decomposing technique	Wavelet transform
ANN Classifier	KNN classifier

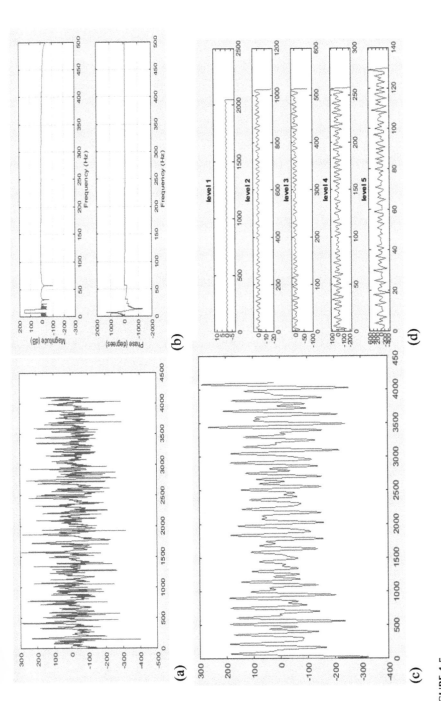

FIGURE 4.5
Qualitative analysis of proposed system on normal EEG signal: (a) input normal EEG signal, (b) magnitude and phase plot of the filter, (c) filtered signal, (d) wavelet decomposition level 5.

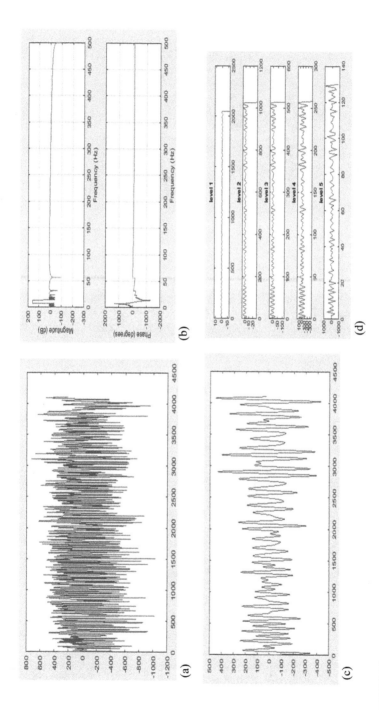

FIGURE 4.6

Qualitative analysis of proposed system on normal EEG signal: (a) input normal EEG signal, (b) magnitude and phase plot of the filter, (c) filtered signal, (d) wavelet decomposition level-5.

TABLE 4.2

Features of Normal and Epileptic EEG Signals

Patient		LPC			Kurtosis	Mean		Auto-Correlation						PCA
1	1.258484	-0.0455	...	-0.0818	624.6202	3.57455	-0.1477	1	...	-0.457	-0.025	0.17166
2	1	1.74397	-0.3164	...	-0.2009	1619.425	3.331812	-0.2513	1	...	-0.164	-0.175	0.20423
3	1	1.367817	-0.1833	...	-0.0946	5079.8	4.538614	-1.9356	1	...	-0.497	0.1293	0.12803
4	1	1.112386	0.102362	...	0.009233	6133.054	2.962493	0.378282	1	...	-0.514	-0.091	0.20748
5	1	1.317612	-0.15653	...	-0.10689	7704.017	2.969006	2.610246	1	...	-0.525	0.1911	0.02850
6	1	1.325386	-0.01242	...	0.009085	2564.211	2.695167	-0.20193	1	...	-0.487	-0.087	0.27810

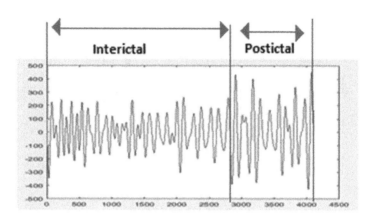

FIGURE 4.7
Seizure activity filtered signal.

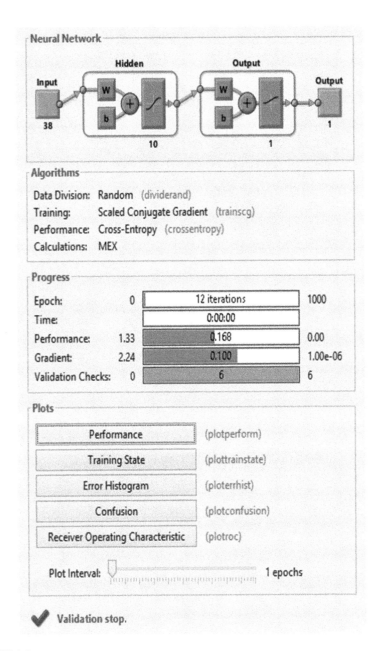

FIGURE 4.8
ANN model.

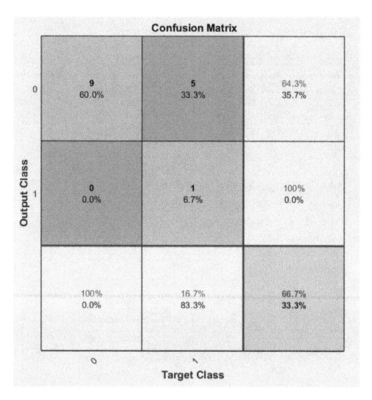

FIGURE 4.9
Confusion matrix.

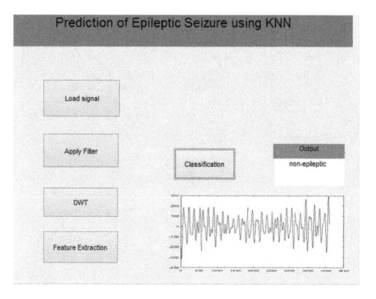

FIGURE 4.10
GUI for non-epileptic patient.

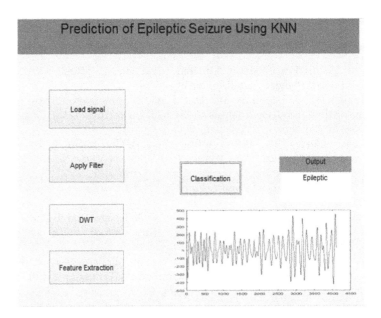

FIGURE 4.11
GUI for an epileptic patient.

4.13 Conclusion

In this chapter, the system for auto-classification of the normal and epileptic EEG signal has been implemented. EEG signals of normal and epileptic patients were collected from online sources. The EEG signals were preprocessed using the Chebyshev filter. From the filtered signal, various features are extracted, viz. LPC, kurtosis, mean, auto-correlation and PCA. Those features are used as an input in a KNN. The final classification of the EEG signals of the existence of seizures or not is done by KNN. A further filtered signal is used to predict the occurrence seizure event. The application can be scaled up using the Web of Things.

4.14 Acknowledgments

We are thankful to Dr. Sanjay Pawar (Fellowship in Stroke Neurology [FISN], Consultant Stroke Physician), for continuous guidance and more insight on epileptic seizure medical terminology.

References

1. World Health Organization. Epilepsy. Available at: https://www.who.int/ news-room/fact-sheets/detail/epilepsy.
2. T. Tzallas, M. G. Tsipouras, and D. I. Fotiadis, "Automatic seizure detection based on time-frequency analysis and artificial neural networks," *Computational Intelligence and Neuroscience*, 2, p. 13, 2007.
3. E. Juárez-Guerra, V. Alarcon-Aquino, and P. Gomez-Gil, *Epilepsy Seizure Detection in EEG Signals Using Wavelet Transforms and Neural Networks*. Springer International Publishing, Switzerland, pp. 261–269, 2015.
4. S. S. Zakareya Lasefr, *Epilepsy Seizure Detection Using EEG Signals*. IEEE, p. 6, 2017.
5. J. Rasekhi, M. R. K. Mollaei, M. Bandarabadi, C. A. Teixeira, and A. Dourado, "Pre-processing effects of 22 linear univariate features on the performance of seizure prediction methods," *Journal of Neuroscience Methods*, 217(1–2), pp. 9–16, 2013.
6. C. A. Teixeira, B. Direito, M. Bandarabadi, et al., "Epileptic seizure predictors based on computational intelligence techniques: A comparative study with 278 patients," *Computer Methods and Programs in Biomedicine*, 114(3), pp. 324–336, 2014.
7. C. Brunner, M. Naeem, R. Leeb, B. Graimann, and G. Pfurtscheller, "Spatial filtering and selection of optimized components in four class motor imagery EEG data using independent components analysis," *Pattern Recognition Letters*, 28(8), pp. 957–964, 2007.
8. M. Bandarabadi, C. A. Teixeira, J. Rasekhi, and A. Dourado, "Epileptic seizure prediction using relative spectral power features," *Clinical Neurophysiology*, 126(2), pp. 237–248, 2015.
9. K. Gadhoumi, J. Lina, and J. Gorman, "Discriminating preictal and interictal states in patients with temporal lobe epilepsy using wavelet analysis of intracerebral EEG," *Clinical Neurophysiology*, 123(10), pp. 1906–1916, 2012.
10. S. Zandi, R. Tafreshi, M. Javidan, and G. A. Dumont, "Predicting epileptic seizures in scalp EEG based on a variational Bayesian Gaussian mixture model of zero-crossing intervals," *IEEE Transactions on Bio-Medical Engineering*, 60(5), pp. 1401–1413, 2013.
11. S. J. Roberts, D. Husmeier, I. Rezek, and W. Penny, "Bayesian approaches to Gaussian mixture modeling," *IEEE Transactions on Pattern Analysis and Machine Intelligence*, 20(11), pp. 1133–1142, 1998.
12. Bahubali Shiragapur, Nishikant Surwade, et al., "Experimental study on detection of epilepsy," *Helix*, 9(3), pp. 5052–5056, 2019.
13. Vijaylaxmi Jain, Bahubali Shiragapur, et al., "Sleep stages classification using wavelet transform & neural network," in *Proceedings of 2012 IEEE-EMBS International Conference on Biomedical and Health Informatics*, Hong Kong, pp. 71–74, 2012.
14. Song Y. Crowcroft and J. Zhang, "Automatic epileptic seizure detection in EEGs based on optimized sample entropy and extreme learning machine," *Journal of Neuroscience Methods*, 210(2), pp. 132–146, 2012.
15. CHB-MIT Scalp EEG Database. *PhysioNet*. Available at: https://physionet.org/ content/chbmit/1.0.0/.

16. Farzad Samie, Sebastian Paul, et al. "Highly efficient and accurate seizure prediction on constrained IoT devices," IEEE Xplorer, 2018, ISSN-1558-1101.

17. Yinda Zhang, S. Yang, et al., "Integration of 24 feature types to accurately detect and predict seizures using scalp EEG signals," *Sensors*, 18(5), p. 1372, 2018. doi:10.3390/s18051372.

18. V. V. Shete, et al., "Sleep stage classification using wavelet transform & neural network," *Trans-Stellar Journal USA, IJECIERD*, 2(2), pp. 38–45, 2012.

19. K-nearest neighbor classifier. *Python Machine Learning Tutorial*. Available at: https://www.python-course.eu/k_nearest_neighbor_classifier.php.

5

Cognitive Radio Networks

Kirti Dalal and Aarti Jain

CONTENTS

5.1 Introduction

The frequency spectrum is considered the most pivotal, yet limited, natural resource. The rapid utilization of wireless technology in communication has escalated the demand for larger bandwidths. The number of users with internet-enabled wireless mobile devices is increasing rapidly, which attracts the proliferating need for more advanced applications. These applications include multimedia communication, telemedicine, smart spaces (e.g. office, home, etc.), sensor networks, smart cities and many more. This requires transferring higher volumes of data and providing higher security, as well as continuous connection among the wireless mobile devices of any network, at any time and at any place.

The present mobile communication technology is based on the cellular concept which uses a "Static Spectrum Allocation Scheme." In a static spectrum allocation scheme, the legacy owner is assigned the entire licensed band which can only be utilized by the primary users. Many of the pre-allocated frequency bands are ironically underutilized since the primary users may not be active at all times. Hence, the resources are simply being wasted. However, Cognitive Radio Technology adopts a "Dynamic Spectrum Allocation Scheme," in which both licensed as well as unlicensed users can use the spectrum efficiently, thus mitigating the spectrum shortage problem and fostering easier and more flexible radio spectrum access to wireless networks. Moreover, Cognitive Radio Networks (CRNs) help to ameliorate the Quality of Service (QoS) by providing congestion control, higher bandwidth and faster data rate.

5.2 Cognitive Radio

Joseph Mitola and Gerald Maguire were the pioneers who introduced the concept of Cognitive Radio (CR). The word "cognitive" appertains to conscious mental activity like hunting, understanding, learning and remembering. A Cognitive Radio is an astute device which can sense, learn and react to the network conditions. In other words, CR is a radio which can revamp its transmission parameters depending upon the environmental conditions under which it functions.

5.2.1 CR Transceiver System

To realize a CRN, each CR node must have a CR transceiver system which could function as a transmitting unit as well as a receiving unit. A basic CR

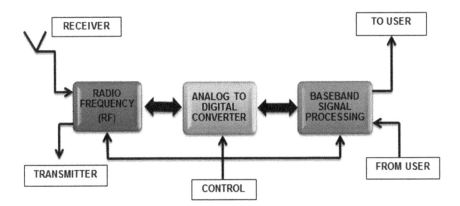

FIGURE 5.1
CR transceiver system.

transceiver system has three units: A radio-frequency (RF) unit, an analog-to-digital (A/D) converter and a baseband signal processing unit. The RF unit along with the A/D converter forms the RF Front End. The basic block diagram of a CR transceiver system is depicted in Figure 5.1.

At first, the acquired signal is amplified by the RF front end and converted into a digital signal. Then the baseband signal processing unit performs modulation and encoding of the digital signal before sending it to the user. Moreover, on the contrary it performs decoding and demodulation of the digital signal after receiving the signal from the user.

5.2.2 Spectrum Hole Concept

The legacy owner has access to the assigned or licensed primary band. However, it may not be using this pre-allocated frequency band at all times, i.e. there would be instances where this band would be left idle by the legacy owner. These vacant or unused primary bands are entitled "white spaces" or "spectrum holes." Spectrum holes are the medium through which CR users sustain communication and perform their normal tasks. A CR user has to optimistically choose the spectrum hole which can meet its QoS and must move to new white space, for its seamless transmission, when the licensed user reappears. The concept of spectrum holes is illustrated in Figure 5.2.

5.2.3 Network Architecture

The modern wireless network models utilize heterogeneous spectrum customs and telecommunication techniques. Furthermore, a part of the existing radio spectrum is accredited to various technologies, hence only few bands are available for free access (e.g. Industrial, Scientific and Medical

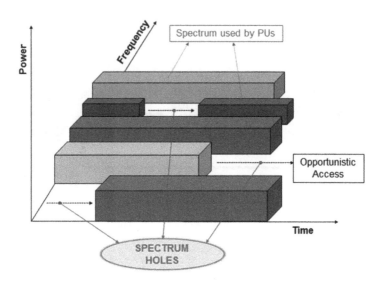

FIGURE 5.2
Spectrum hole concept.

[ISM] band). A set of rules, called protocols, are required for communication among the CRNs. For the development of such communication protocols, a comprehensible illustration of network architecture is required.

The elements of network architecture, as illustrated in Figure 5.3, could be categorized into two subsets, that is, the primary network and the cognitive radio network. The components of these two subsets are discussed below.

5.2.3.1 Primary Network

The network which has the license to access a particular band of the radio spectrum is called the Primary Network (PN). For instance, a PN could be the cellular network, CDMA, WiMAX or a TV broadcast network. A PN consists of the following components:

1. Primary User (PU): The user of a PN which has the rights to utilize the particular spectrum band allocated to PN is called the PU. The PU can access the PN via base station. All the services and operations of the PU are controlled by the base station. The PU is the matter of greatest importance; hence it has precedence over the CR user when it comes to channel access or channel allocation. Thus, any unlicensed user or user of any other network should not affect the PU. A PU doesn't require any changes in its parameters for coexistence with the CR base stations and CR users.

2. Primary Base Station (PBS): PBS is the permanent infrastructure network element deployed for a particular technology with authorized

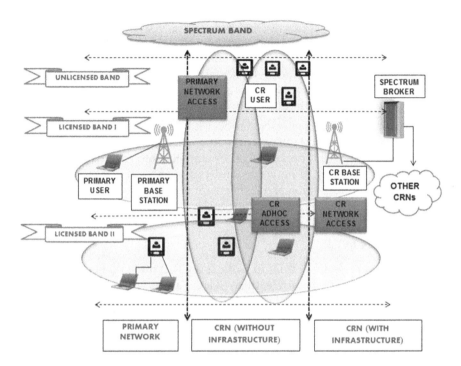

FIGURE 5.3
Network architecture.

access to a spectrum band. For instance, a Base-station Transceiver System (BTS) of a cellular network or WiMAX are the fixed infrastructure element of the network model. PBS is incapable of coexistence with CRN and thus a few changes are required in PBS for the cooperation of CR users in the PN. These changes could comprise the addition of CR protocols, required for the access to PN, in the PBS along with the protocols of PU.

5.2.3.2 Cognitive Radio Network

The network which doesn't have the authority or rights to function in a particular band but opportunistically gains access to the spectrum is called a CRN. A CRN could be an infrastructure network or an ad hoc network, as illustrated in Figure 5.3. A CRN has the following components:

1. CR User: A CR user or secondary user (SU) is the user which has no spectrum license for its operation. The SU can gain access to the channel only when the PU is absent or not active. Hence, the SU has to desert the occupied channel whenever the PU comes back without provoking any interference to the PU.

2. CR Base Station: CR users could access the spectrum either via a non-infrastructure-based network (ad hoc access) or via an infrastructure-based network. A CR base station or secondary base station is a permanent infrastructure component that provides a single-hop connection to CR users without any license for the radio spectrum. A CR user could gain access to the other networks with the help of this connection.

3. Spectrum Broker: A spectrum broker is a central network entity that governs the sharing of spectrum resources among different CRNs. Hence, a spectrum broker could be connected to each network like the star topology in networks and could act as centralized server having all information about spectrum resources required to enable multiple CRNs to exist simultaneously.

5.2.4 Features of CR

For any software-defined radio to work as a CR, it must possess some basic characteristics. The two main features of a CR are as follows:

1. Cognitive Capability: This is the propensity of a CR user (SU) to discern and assemble the details like bandwidth, transmission frequency, modulation and the power of the PU from the radio environment.

2. Reconfigurability: This is the potentiality of CR user (SU) to rectify its specifications like transmission power, operating frequency and modulation depending upon the collected data without any alteration in the hardware components.

5.2.5 Cognitive Cycle

A CR follows a cognitive cycle in order to recognize and exploit the unused spectrum channels or holes whenever they are vacated by the PU. This is illustrated in Figure 5.4. The basic functions performed by a CR are as follows:

1. Spectrum Sensing: The CR has to sense the radio spectrum, determine the spectrum holes or white spaces and encapsulate their information.

2. Spectrum Decision: The CR has to select the optimum spectrum hole among all the sensed white spaces and establish the transmission parameters for the SU.

3. Spectrum Sharing: The CR has to administer the spectrum access among all the CR users for efficient utilization of the spectrum.

4. Spectrum Mobility: The CR has to vacate the spectrum whenever the PU shows up and shift to another white space for its continuous transmission.

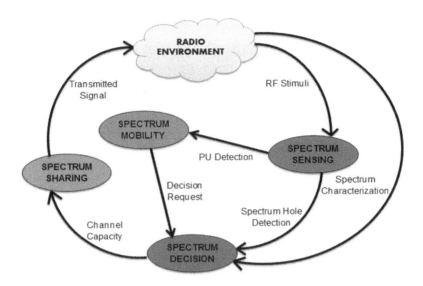

FIGURE 5.4
Cognitive cycle.

5.3 Spectrum Sensing

The prime agenda of CR is to utilize the white spaces wisely. Apart from sensing the idle frequency band of a PU that could be allocated to an SU, the SU should also be able to sense the advent of a PU and shift itself to some other white space for its seamless transmission without offering any interference to the PU. Hence, sensing schemes play a crucial role in CRNs.

There are various kinds of spectrum sensing schemes used to exploit the detection of spectrum holes. The classification of spectrum sensing schemes can be seen in Figure 5.5.

5.3.1 Non-Cooperative Sensing

Non-Cooperative Spectrum Sensing (NCSS) techniques are signal processing techniques in which the existence of a PU in a specific spectrum is independently decided by each CR. They are further divided into following three types.

5.3.1.1 Matched Filter Detection

The Matched Filter Detection (MFD) technique is a coherent detection technique. The PU's signal information, like operating frequency, pilots, spreading codes, modulation technique, preambles, packet format, transmission

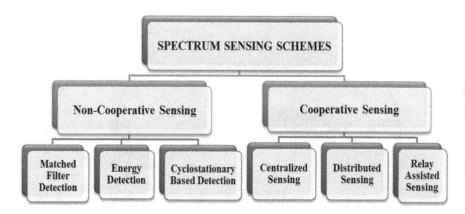

FIGURE 5.5
Classification of spectrum sensing schemes.

power, and so on, are required beforehand, that is, the CR user must have the knowledge about all the transmission parameters of the PU. In order to detect the existence of the PU, sensing is performed by correlating the observed signal and the known sample.

The MF operation is similar to correlation where the PU or the unknown signal is convolved with a filter having the impulse response as the mirror and a time-shifted version of the reference signal for maximizing the output signal-to-noise ratio (SNR). In MFD, the input is passed through a Bandpass Filter (BPF), after which it is convolved with the MF having an impulse response that is the same as the reference signal. The output of the MF is then compared with a threshold for decision-making as depicted in Figure 5.6. The MF is modeled from the given Equation (5.1):

$$y(n) = \sum_{k=-\infty}^{\infty} h(n-k)x(k) \tag{5.1}$$

where x is the unknown signal vector that is convolved with the impulse response h of the filter, and y is the output.

Detection using MF is very fast because it entails only few samples to meet up with a specified probability of detection constraint. When the SNR is low, the number of samples needed is of the order $1/(SNR)$, in contrast to energy detection in which the required number of samples is of the order $1/(SNR)^2$.

FIGURE 5.6
Matched filter detection.

However, MF demodulates PU signals, hence CR requires perfect knowledge of PU signal. Also its performance is poor in the event of frequency or time offset. The complexity of MF is further increased, since CR needs to have a staunch receptor for all types of PU signals for effective detection. MF is not efficient in terms of consumption of power or energy because many receiver algorithms have to be run for all types of PU signals.

5.3.1.2 Energy Detection

The Energy Detection (ED) technique, unlike MFD, is a non-coherent detection technique where no prior information about the PU signal is required. The ED tactic is the most commonly used spectrum sensing technique owing the fact that the computational cost is low.

PUs are flexible and agile for selecting their modulation type and pulse shaping schemes which may not be known to the CR users. In such situations where the PU signal structure is not known to the SU, the ED technique is the optimal spectrum sensing technique. In this, the power of the signal in a given channel is calculated and is compared with a predefined threshold for decision making.

ED is based on the test of binary hypotheses given in Equation (5.2). The test statistics for ED are given in Equation (5.3):

$$0: y(n) = w(n)$$

$$1: y(n) = hx(n) + w(n) \tag{5.2}$$

where $w(n)$ is the additive white Gaussian noise (AWGN), $x(n)$ is the PU signal at time n, h is the channel gain and $y(n)$ is the measured signal, while 0 and 1 are the two hypotheses representing that the PU is absent or present, respectively.

$$T = \sum_{k=1}^{N} (y(k))^2 \tag{5.3}$$

In ED, if the test statistics are greater than the set threshold, the PU is assumed to be active or present, otherwise the channel is vacant. The choice of threshold depends on the relative cost of false alarms and missed detection.

The time domain ED consists of a channel filter to eliminate the band of noise and the adjacent signals, a Nyquist sampling analog-to-digital (A/D) converter, a square law device and an averaging unit as depicted in Figure 5.7. ED could also be implemented in frequency domain by performing an averaging operation on the frequency bins of the fast Fourier transform (FFT), as depicted in Figure 5.8.

The major advantage of ED lies in the simplicity of the algorithm and the attribute that it does not need any prior information about the type of PU

FIGURE 5.7
Energy detection in time domain.

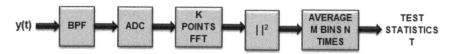

FIGURE 5.8
Energy detection in frequency domain.

signal; rather it only requires the knowledge of the noise parameters. The main disadvantage of ED is that it is not robust when the noise variance is not known or is time variant. Also the performance is very poor when the signal to SNR is very low because the detector can't differentiate between the PU signal and the SU signal or noise, since the PU signal is modeled as a zero-mean white stationary Gaussian process.

5.3.1.3 Cyclostationary-Based Detection

In the Cyclostationary-Based Detection (CBD) technique, the existence of PUs periodicity in the received PU signals needs to be discovered. The periodicity is basically ingrained in pulse trains, sinusoidal carriers and spread spectrum sequences (i.e. direct sequence spread spectrum [DSSS] codes and frequency-hopped spread spectrum [FHSS] codes) of the PU signals. These signals demonstrate characteristics of periodic statistics and spectral correlation which distinguish them from noise and interferences. A signal is said to exhibit cyclostationarity if, and only if, the signal is correlated with certain frequency-shifted versions of itself. If the cyclic frequency is equal to the fundamental frequencies of the transmitted signals, the cyclic spectral density (CSD) gives the peak values.

The prime merit of CBD is that it can distinguish the PU signals from noise and can detect PU signals with low SNR. Its main demerit is that the algorithm is complex with long observation or sensing time. Moreover, it needs prior knowledge about the PU signal that may not be accessible to CRs. Because of its high complexity and slow detection speed, this spectrum sensing scheme is rarely used.

The received signal is given by Equation (5.4), while the CSD function is as shown in Equation (5.5). Figure 5.9 illustrates the basic block diagram of CBD.

$$y(n) = h\, x(n) + w(n) \tag{5.4}$$

where $w(n)$ is the additive white Gaussian noise (AWGN), $x(n)$ is the PU signal at time n, h is the channel gain and $y(n)$ is the measured signal.

$$R_y^\alpha(\tau) = E\left[y(n+\tau)y^*(n-\tau)e^{j2\pi fn}\right]$$

$$S(f,\alpha) = \sum_{\tau=-\infty}^{\infty} R_y^\alpha(\tau)e^{-j2\pi ft} \qquad (5.5)$$

where $R(\tau)$ is a cyclic auto-correlation function, f and α are the fundamental signal frequency and cyclic frequency, respectively, and τ is time shift.

All the spectrum sensing schemes discussed above have their own pros and cons. Since they are the most important techniques in CRNs, the relative analysis of all the NCSS schemes is presented in Table 5.1.

5.3.2 Cooperative Sensing

In cooperative spectrum sensing (CSS), the spectrum sensing task is carried out by multiple CRs working collaboratively. This approach aims at solving spectrum sensing problems resulting from noise uncertainty, fading, shadowing and faulty sensor of a single detector. The reliability of detecting a weak primary signal using one CR is difficult to maintain because of poor

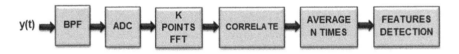

FIGURE 5.9
Cyclostationary-based detection.

TABLE 5.1
Relative Analysis of the NCSS Schemes

Scheme	Merits	Demerits
Matched Filter Detection	• Requires less sensing time and a lower number of samples. • Provides best SNR.	• A prior knowledge about the PU's characteristics is required. • Consumes more power. • High-computational complexity.
Energy Detection	• Computational complexity is low. • Doesn't require prior information of PU's signals.	• Provides poor SNR. • Requires long sensing time for accurate results.
Cyclostationary-Based Detection	• More resilient to noise levels, thus provides better results at low SNR.	• Highly complex. • Requires prior information about the PU's characteristics. • Can't provide high detection speed.

conditions of channels due to multipath fading and shadowing between the PUs and the SUs. Users at different distant geographical locations experience an independent fading condition. The effects of destructive channel conditions resulting from fading, shadowing or hidden nodes can be mitigated if the users operate cooperatively. This will increase the overall detection reliability. When cooperation is allowed among users, individual SUs require less sensitivity to achieve high detection reliability. CSS is found to decrease significantly both the probability of false alarm (Pf) and the probability of miss detection (Pm), and simultaneously offers more protection to PUs. It also decreases the sensing time as compared with single local sensing methods.

CSS makes use of the advantages of spatial diversities of CRs to enhance the probability of detection (Pd). CSS comprises a control channel that is used for communicating spectrum sensing results and channel allocation information among CR users. The main challenge in CSS is the development of efficient spectrum information-sharing algorithms and the complexity of the approach. The ED technique is the current approved detection technique for use in CSS due to its non-coherency and simplicity.

There are three major topologies proposed for achieving CSS in CRNs according to their level of cooperation. They are discussed below.

5.3.2.1 Centralized Cooperative Sensing

In Centralized Cooperative Sensing (CCS), there is a central entity named a server or cluster head or Fusion Center (FC), which gathers the sensing data from the cognitive devices, locates the unutilized spectrum and sends this data to other CR users, or personally administers the CR traffic. FC could be a wired mobile device, access point, base station or another CR.

There are certain steps that have to be followed in CCS. First, the FC has to select a range of frequency or a channel to perform the sensing operation. It then orders every cooperating CR to perform sensing personally. Second, all cooperating CRs have to divulge their sensing results to FC. At last, the FC needs to amalgamate the received sensing information and make a verdict about the existence of PU. The FC not only disseminates the diffused verdict back to the CRs but also ensures that the available white spaces are shared efficiently among all CRs. This scheme is illustrated in Figure 5.10(a).

CCS schemes could further be grouped into two main classes: partially and totally cooperative schemes. In totally cooperative schemes, CRs cooperatively sense the channels and also relay each other information in a cooperative way. In partially cooperative schemes, cooperation is limited to sensing the channels, while the CRs detect the channels independently and communicate their results and decision to the FC.

The objective is to alleviate the effects of fading on the channel and boost the performance of detection. A dedicated link between FC and CRs is required to exist all the time in this technique. This increases the cost of the system.

5.3.2.2 Distributed Cooperative Sensing

In Distributed Cooperative Sensing (DCS), the FC is not present, hence each cluster node individually senses the spectrum and shares the gathered intra-cluster information among all other cluster nodes. Later, each cluster node individually decides which spectrum hole it can use, keeping in mind all the information gathered from other CR users as well. This scheme is illustrated in Figure 5.10(b).

The advantage of DCS over CCS is that since extra infrastructure is not required, the cost of the network is reduced. However, due to the absence of an FC, this scheme needs perpetual upgrading of the table containing the information about the spectrum which demands huge storage capacity and computational skills. The clustering or gossiping algorithm is one of the algorithms suggested to be adopted by the DCS scheme.

5.3.2.3 Relay-Assisted Cooperative Sensing

In Relay-Assisted Cooperative Sensing (RACS), the information is shared among CR users in a decentralized method. This scheme is used when the forward channel (sensing channel) and reverse channel (reporting channel) are not properly synchronized. Each CR user individually identifies the white spaces in the spectrum and independently makes the decision. When the PU reappears, it immediately relinquishes the channel without notifying the other users.

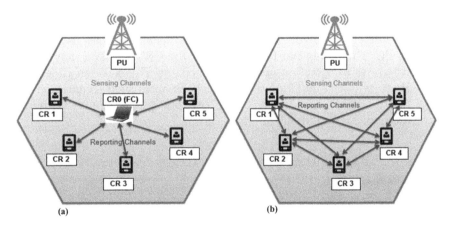

FIGURE 5.10
(a) Centralized cooperative sensing. (b) Distributed cooperative sensing.

Though the time taken for detection under this scheme is small, it needs dedicated hardware for cooperation, which escalates the overall cost of the system. This technique is very weak and inefficient in maximizing the usage of the limited spectrum resources. It also prompts causing interference to PU as a result of miss detection.

5.3.3 Spectrum Sensing Issues

There are several hurdles that may arise while performing spectrum sensing in CRNs which still need to be tackled. The most frequent problems are:

1. Hidden PUs: Various factors, such as shadowing and multipath fading, cause hidden user problems. This problem could be handled by using CSS schemes but still requires improvement for positive results.

2. Detection of Spread Spectrum Users: The technologies used in commercial devices are of two types, i.e. fixed frequency spectrum and spread spectrum. Spread spectrum technique can again be of two types, i.e. FHSS and DSSS. When a PU uses spread spectrum technique, it becomes difficult to find the exact transmission power of the PU, since the power gets distributed over a vast range of frequencies. Moreover, when the pattern of hopping is not known in FHSS, this hurdle becomes unavoidable and absolute synchronization can never be obtained.

3. Sensing Duration and Frequency: Any spectrum sensing scheme must be capable of detecting the existence of the PU within a reliable amount of time in order to avoid interference. For proper synchronization between the sensing duration and reliability of sensing, it is important to adopt proper sensing parameters such as channel move time, channel detection time, sensing frequency and many more.

4. Security: Any detrimental node can alter its air interface to imitate as PU. This could not be detected by the CR users and may lead to wrong information about the spectrum. This is often called a Primary User Emulation (PUE) attack. To avoid such scenarios and to strengthen the security of the network, public key encryption methods are required.

5.4 Spectrum Decision

Once the spectrum is sensed, the CRN must make a decision about the optimum spectrum hole which could provide the desired QoS. Spectrum decision consists of three factors: spectrum characterization, spectrum selection and spectrum reconfiguration.

At first, the spectrum is characterized based upon the data gathered through spectrum sensing. Second, spectrum selection is achieved by choosing the spectrum hole which meets the desired QoS. At last, the transmission parameters like power, bandwidth, and so on, are reconfigured.

5.4.1 Spectrum Decision Parameters

There are certain parameters which need to be kept in mind while arriving at a spectrum decision, because they directly influence the QoS provided to the users. They are as follows:

1. Interference: A CR node must transmit within the maximum allowable interference level. This allowable limit of interference is used to compute the power of a CR node. In order to decrease the interference caused to the PU and to increase the capacity of the channel, proper power allocation is required.

2. Path Loss: Path loss is directly proportional to the frequency and distance. When the frequency of operation is increased, path loss also gets increased which decreases the range of transmission. Path loss could be reduced by increasing the transmission power; however, this would lead to interference among other users.

3. Link Layer Delay: The CR users operating in a network may work on different types of link layer protocols. These link layer protocols may have varying parameters which may result in different link layer delays.

5.4.2 Spectrum Decision Issues

There are many issues that arise while making the spectrum decision. A few of them are discussed below:

1. PU Activity Modeling: Many specifications of PU cannot be simply modeled by using the simple ON or OFF model as considered under binary hypotheses. Inaccurate models adversely affect the spectrum decision.

2. Joint Spectrum Decision and Reconfiguration: Even after the spectrum is chosen by evaluating the desired QoS, the spectrum specifications may vary with time. Hence, an ideal combination of spectrum selection as well as reconfiguration is required for efficient utilization of the spectrum.

3. Information about the Location: To determine the interference level, the CR user must know the exact location and the transmission power of PU. However, this information may not always be available to the CR user.

5.5 Spectrum Sharing

Spectrum Sharing shows how the spectrum is being distributed among the CR users. It administers the access of channel to the CR users so that the desired QoS can be maintained without provoking any interference to the PU. The spectrum sharing includes both the techniques of sharing spectrums among multiple CRNs, as well as techniques of sharing spectrums within a CRN.

5.5.1 Spectrum Sharing Techniques

The elucidation of spectrum sharing in CRN could be broadly classified into three aspects: Network Architecture, Access Technology and Allocation Behavior.

According to the Network Architecture, spectrum sharing could be of two types:

1. Centralized Spectrum Sharing: Under this, the spectrum access and allocation is controlled by a central entity. This central entity builds the spectrum allocation map according to the information provided by each user in the network according to their requirements.
2. Distributed Spectrum Sharing: Under this, every user is independently in charge of its spectrum access and the allocation according to the local policies and no central entity is needed. When construction of infrastructure is not possible, this spectrum sharing technique is used.

According to the Access Technology, spectrum sharing could be of two types:

1. Underlay Spectrum Sharing: Under this, CR users and PU can simultaneously transmit the data but no interference should be caused to the PU. To achieve this, CR users must maintain their transmission power lower than a threshold value to curtail the interference with PU.
2. Overlay Spectrum Sharing: Under this, CR users can utilize the spectrum only when the PU is not present or inactive. Hence, the interference with PU is less, but spectrum utilization is minimum.

Based on the Allocation Behavior, spectrum sharing could be of two types:

1. Unlicensed Spectrum Sharing: Under this, no user has any predominance, i.e. all the users are treated equally. When this unlicensed spectrum is vacant any CR user can acquire access to it.

2. Licensed Spectrum Sharing: Under this, access to spectrum depends on the primacy. PUs are kept at a higher position in the preference table for spectrum allocation. CR users are allocated the spectrum only when the PU is absent.

5.5.2 Spectrum Sharing Issues

There are several issues which may arise while sharing the spectrum among the different users. A few of these are discussed below:

1. Distributed Power Measurement: In a distributed spectrum sharing scheme, the transmitting power of a CR user is measured in the distributed manner, since there is no central entity. Hence, complex power control methods are required.
2. Discovering the Topology: Since different CR users use different spectrum holes, i.e. non-uniform channel allocation, it becomes difficult to find the actual topology of the network.

5.6 Spectrum Mobility

The CR user has to abandon its current spectrum as soon as the PU reappears at that specific spectrum location. For a CR user to continue its communication without any interruption, it has to shift to another white space which could provide the desired QoS. This shifting is called Spectrum Mobility (SM). SM could also occur due to link failure, i.e. rupturing of the communication link. The prime task of SM is "Spectrum Handoff" or "Spectrum Handover."

5.6.1 Spectrum Handover Strategies

Different handover strategies could be used in CRN for SM. Some of them are discussed below:

1. Non-Handover Strategy: In this method, the SU stays ineffective until the channel becomes vacant again. This means that the next target channel to which the SU has to move is the current channel of the SU where it was operating. The prime demerit of this strategy is the wastage of time of the SU due to high waiting latency. Its merit is the low PU interference. This strategy is applicable for short data transmission.

2. Pure Reactive Handover Strategy: In this method, the SU moves to another spectrum hole after the detection of link failure. This means that the SU solicits reactive spectrum sensing and reactive handover action. The key demerit of this strategy is the delay in spectrum sensing, but accurate channel selection is its merit. This strategy is well suited to short sensing time data.

3. Pure Proactive Handover Strategy: In this method, the SU moves to another spectrum hole before the detection of link failure. This means that the SU solicits proactive spectrum sensing and proactive handover action. The merit of this strategy is the low handover latency because it could formulate all the parameters beforehand. However, the comprehensive performance of SM could be deteriorated by poor spectrum sensing. This strategy is well-suited to large time-sensing data.

4. Hybrid Handover Strategy: This method is the combination of pure reactive strategy and pure proactive strategy. Although the SU already determines the next target channel, it moves to the new channel only after the occurrence of handover triggering. Hence, the spectrum handover time is very fast. This strategy is suitable for all basic PU networks.

5.6.2 Spectrum Mobility Issues

Various problems may arise while performing SM in order to maintain the desired QoS and provide seamless connectivity. Some of these problems are discussed below:

1. SM in Time Domain: SU could choose the white space only for its present transmission based on the requirements of QoS. However, with the passage of time, the available spectrum also changes which may not be appropriate to provide the desired QoS. Hence, perpetuating the same desired requirements of QoS throughout the entire transmission becomes onerous.

2. SM in Space Domain: The SU has to move from one location to the other whenever the PU reappears. Hence, the available spectrum also gets changed every time, which makes the perpetual spectrum allocation a difficult task.

3. Energy Efficiency: SM strategies relay the data gathered through spectrum sensing. Insufficient information about the spectrum leads to reduction of energy efficiency of the network, thus making SM arduous.

4. Switching Delay: When the SU moves from one spectrum location to the other location, the time taken to switch the spectrum should be minimal, or else the data transmission will fail.

5. Adaptive Spectrum Handover Strategy: The most appropriate handover strategy should be selected according to the PU traffic pattern. Moreover, the SU must be able to adapt to a new appropriate handover strategy whenever this PU traffic pattern becomes varied.

5.7 Applications of CRNs

CRN is a technology that it is hoped will mitigate the issue of spectrum scarcity, spectrum congestion and Quality of Service (QoS). This technology is still in its embryonic stage and there is still much to be explored in the coming future. Some of the fields of application of CRN include:

1. Leased Network: PN may provide a leased network by permitting CR users to gain access to its licensed spectrum in an opportunistic manner without harming the communication of the PU.

2. Cognitive Mesh Network: For providing broadband connectivity, wireless mesh networks are coming up as a cost-effective technology. However, mesh networks need a higher capacity to meet the requirements of the applications that demand higher throughput. Since the CR technology facilitates the access to larger amount of spectrum, CRNs will therefore be an adequate option to meet the requisites of mesh networks.

3. Emergency Network: CRNs can be implemented for public safety and emergency networks. Under the situations of natural disasters where PNs are temporarily distressed, their spectrum band can be used by CR users. CRNs can establish communication on the available spectrum band in ad hoc mode without the need for an infrastructure and by maintaining communication priority and response time.

4. Military Network: CRNs can be used in a military radio environment. CRNs can enable military radios to select arbitrary intermediate frequency (IF) bandwidth, coding schemes and modulation schemes, adjusting to the variable radio environment of the battlefield.

5. Wireless Sensor Network: The traditional Wireless Sensor Network (WSN) operates in unlicensed bands like ISM band which are heavily

congested. If the WSN is embedded with cognitive capabilities, it will provide new dimensions and opportunities to researchers and industry that would assist in designing new algorithms, hardware and software. This would ultimately help to minimize collision, latency and interference by efficient channel utilization, thus reducing the power consumption and increasing the network lifetime.

5.8 Conclusion

This chapter focused on the "Cognitive Radios" in "Mobile Networks." It started with the basic introduction of a CRN, illustrating the concept of spectrum holes, describing the CRN architecture model and explaining the features of a CR. The four key elements of a Cognitive Cycle: spectrum sensing, spectrum decision, spectrum sharing and spectrum mobility, were introduced.

The chapter explained the various Spectrum Sensing schemes used in the CRNs. They were broadly classified as NCSS schemes and CSS schemes. The three types of NCSS schemes, namely MFD, ED and CBD were discussed in detail with their block diagrams and mathematical interpretations. Further, the three types of CSS schemes, namely CCS, DCS and RACS were described. Later, the parameters of Spectrum Decision, like path loss, interference and link layer delay, were discussed. Different types of Spectrum Sharing techniques according to the Network Architecture, Access Technology and Resource Allocation were described. Finally, Spectrum Mobility was discussed and different types of handover strategies like the Non-Handover Strategy, the Pure Reactive Handover Strategy, the Pure Proactive Handover Strategy and the Hybrid Handover Strategy were explained. The challenges and issues arising in all these steps were also mentioned simultaneously.

CRN is an technology that is hoped will mitigate the issue of spectrum scarcity, spectrum congestion and QoS. It finds useful deployments in the military, home appliances, WSN, real-time surveillance, healthcare, vehicular networks and many more fields. This technology is still in its embryonic stage and there is still much to be explored in the coming future.

Bibliography

J. Mitola, "Cognitive radio: An integrated agent architecture for software defined radio", PhD thesis, KTH Royal Institute of Technology, Stockholm, Sweden, 2000.

Niranjan Muchandi and Rajashri Khanai, "Cognitive radio spectrum sensing: A survey", *International Conference on Electrical, Electronics, and Optimization Techniques (ICEEOT)*, 2016.

Simon Haykin, "Cognitive radio: Brain-empowered wireless communications", *IEEE Journal on Selected Areas in Communications*, 23(2), pp. 201–220, 2005.

Joshua Abolarinwa, Nurul Mu'azzah, Abdul Latiff, Sharifah Kamilah, Syed Yusuf and Norsheila Fisal, "Energy efficient, learning inspired channel decision and access technique for cognitive radio based wireless sensor networks", *International Journal of Multimedia and Ubiquitous Engineering*, 10(2), pp. 11–24, 2015.

Akhila Asokan and R. AyyapaDas, "Survey on cognitive radio and cognitive radio sensor networks", *IEEE Conference on Electronics and Communication Systems*, pp. 1–7, 2014.

Amir Ghasemi and Sousa S. Elvino, "Spectrum sensing in cognitive radio networks: Requirements, challenges and design tradeoffs", *IEEE Communications Magazine*, April 2008.

V. Amrutha and K. V. Karthikeyan, "Spectrum sensing methodologies in cognitive radio networks: A survey", *Proceedings of the IEEE International Conference on Innovations in Electrical, Electronics, Instrumentation and Media Technology (ICIEEIMT)*, 2017.

B. Wang and K. J. R. Liu, "Advances in cognitive radio networks: A survey", *IEEE Journal of Selected Topics in Signal Processing*, 5(1), pp. 5–23, 2011.

T. Yucek and H. Arslan, "A survey of spectrum sensing algorithms for cognitive radio applications", *IEEE Communications Surveys and Tutorials*, 11(1), pp. 116–130, 2009.

E. Axell, E. G. Leus, G. Larsson and H. V. Poor, "Spectrum sensing for cognitive radio: State-of-the-art and recent advances", *IEEE Signal Processing Magazine*, 29(3), pp. 101–116, 2012.

S. M. Mishra, A. Sahai and R. W. Brodersen, "Cooperative sensing among cognitive radios", *Proceedings of the IEEE International Conference on Communications*, pp. 1658–1663, June 2006.

6

Extended Paradigms for Botnets with WoT Applications: A Review

Manju Khari, Renu Dalal and Pratibha Rohilla

CONTENTS

During the previous decade, notable factual attempts have been made in the creation of statistical methods that could provide potent and efficient botnet detection, prevention and mitigation. As a consequence, a range of methods for detection based on various technological principles and facts and focusing on various facets of botnet life in general have been discovered. Network traffic is the key principle for discovering botnet's existence, because they pivot around the internet for communicating information with invaders and furthermore for administering various items of attack propaganda. Many ultra-modern pathways use machine learning methodologies and techniques for discovering aggressive traffic. This chapter presents a brief study of the botnet life cycle and modern detection methods for identifying botnet network traffic.

6.1 Botnet Introduction

Classically, botnets emanated from a text-based chat system called Internet Relay Chat (IRC) that systemizes communication in channels. Botnets primarily rely on supervision of communication in IRC chatrooms. IRC chatrooms provide various features like support for administration, games, logins, file sharing, texting, tracking last seen times, email addresses, aliases, and so on. Eggdrop was the first popular IRC bot written in the C language and was recognized in 1993 and thereafter developed. All known IRC bots appeared to have evolved from the same intrinsic idea, but motivation behind these bots is to take over other IRC users or even all servers. Some IRC bots are EFNet, IRCNet, QuakeNet, and so on. These bots were also used to implement malware propagation, Denial of Service (DoS) and Distributed Denial of Service (DDoS) attacks.

Newly evolved bots use complicated procedures for establishing links with other bots and the botmaster, which exploits accessible internet communication protocols and commingles modern, powerful, effective form of circumvention, so that the bot becomes more advanced, complex and sturdy. They hide like viruses and walk through like worms to initiate coordinated offense. Some examples are Sality, Conficker, AgoBot and SDBot. Intrusion Detection System (IDS) software is useful in monitoring network activity and loitering for abnormal events to arise. Intrusions can be described as ventures to interfere with integrity, confidentiality, availability of data, or to bypass the security procedures of a system. A software application called Bot runs via worms, Trojans or other malicious codes is used to execute a range of cyber functions [2]. A connected group of bots form a network called botnet, which acts on behalf of a human operator called a botmaster. So, a botnet is a system of connected machines on a network that are infected, frequently called zombies, and are contaminated with malware that allows an attacker to rule them. Every PC in a botnet is called a bot. One who takes control of botnets is a botmaster. Botnets are commonly employed for click fraud, spamming or DDoS) attacks. Botnet circumventions are ubiquitous. Mostly intrusion detection centers on individual hosts don't focus on preventing botnet shaping. Botnets use IRC, HTTP, and so on, protocols for communication. Here botnets are regarded as infected machines. In these situations, many botnet detection methods came into existence. Some methods include handling botnets using flexible C&C channels. Botnets results in DDoS attacks that aim to prevent normal communication by damaging the resources or the infrastructure that is used for connectivity.

Around 2006, to prevent detection, some botnets were scaled back in size. Some botnets are presented in Table 6.1.

TABLE 6.1

History of Botnets

Serial No.	Date of Creation	Name	Estimated No. of Bots	Target Features
1.	2016	Mirai	380,000	• Infected Linux PCs (personal computers). • Targets WoT (Web of Things) devices like IP cameras, etc.
2.	2013	Zer0n3t	200+ server computers	
3.	2012	Chameleon	120,000	• Infected Windows PCs.
4.	2011	Ramnit	3,000,000	• Infected Windows PCs. • Targets removable drives, e.g. USB flash drives and also hides in the master boot record (MBR). • Not a worm. • Acts as a backdoor.
5.	2010	Kelihos	300,000+	• Involved in email spamming and bitcoin theft.
6.	2010	Zeus	11,000+ 3,600,000	• Trojan horse malware. • Involved in stealing bank credentials, email and social credentials by website browsing, keystroke logging, spam messaging and form grabbing. • Used in installing CryptoLocker ransomware.
7.	2009 (Aug)	Festi	250,000	• Involved in email spam (2.5 million spam emails per day) and denial of service attacks.
8.	2009 (May)	BredoLab	30,000,000	• Involved in viral email spam.
9.	2008	Sality Conficker	1,000,000 10,500,000+	Sality: • Targets files on Microsoft Windows systems. • Involved in spam relaying, communications proxying, sensitive data exfiltration, compromising web servers. Conficker: • Targets files on Microsoft Windows operating system (OS). • Uses vulnerabilities in Windows OS software. • Involved in dictionary attack on administrator passwords.
10.	2007	Cutwail Akbot	1,500,000 1,300,000	Cutwail: • Involved in sending spam emails. • Targets computers running Microsoft Windows. Akbot: • IRC (internet relay chat)-controlled backdoor program. • Used to gather data, kill processes or perform DDOS attacks.
11.	2006	Rustock	150,000	• Targets computers running Microsoft Windows. • Average of 192 spam messages per compromised machine per minute sent.

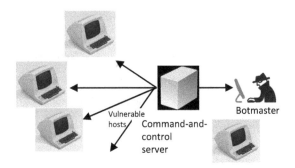

FIGURE 6.1
Botnet components.

6.1.1 Botnet Components

For clear knowledge of how a botnet runs, its basic elements should be understood first. However, there are botnets which follow different structures in order to avoid detection [1,7], as shown in Figure 6.1. A bot is installed malware in an unguarded host which is responsible for performing series of malicious tasks. Installation of this malware can be done through numerous processes, like contaminated websites. These are particularly implemented in such a way that whenever the victim starts their internet-connected machines, only then does the bot initiates its processing. Using a secure command-and-control (C&C) channel, the botmaster sends commands. The main point to note is that bots are not systems or applications weaknesses, but are malware that are spread by contaminated websites, and so on. A botnet is a network of compromised machines called bots and an owner who controls the bot through the command-and-control server to execute malicious activities. Botmasters publish commands to the bots to execute unlawful tasks.

Vulnerable machines are ones on the internet that have been affected with malicious software dispersed by an attacker through various distribution mechanisms. After infection, these hosts become "robots" and can be used as an attacking weapon to carry out many unlawful activities like denial of service (DoS) attacks against other vulnerable hosts. The prime constituent of a botnet is the C&C structure, which consists of compromised hosts called bots and an attacker machine called a controlling station, which can be a centralized or decentralized type [6]. Botmasters use many communication protocols to give directions to their slave nodes and harmonize their actions.

6.1.1.1 Desired Properties of a Bot (Host)

A bot is an internet-connected and unguarded machine on which malicious software is installed to direct a chain of malicious tasks. Installation of such malware can be done in many different ways, which includes various mechanisms or by using contaminated sites. Simply, one could assume that bots will assemble themselves as per different network topologies like bus, star or mesh. These topologies are worthwhile for conventional case study of discrete network characteristics, but don't discusses the vigorous nature of full-size botnets. Rather there are three significant metrical units of botnets: Network diameter and size of targeted network. For example, attackers may want target machines with high bandwidth [2].

Size means the "huge" part of the botnet, or biggest tied (or online) component of the graph. The giant component allows directly counting the deterioration likely caused by certain botnet roles. In the Distributed DOS (DDoS) scenario, the gigantic component S enables us to evaluate the maximum number of bots that can accept instructions and engage in an attack. This opposes the community of all corrupted victims, which perhaps may not be in scope of the botmaster all the time. By network diameter, we mean the mean network's geodesical length. It means average of the minimum distance between each and every connecting node in the network. For the largest network, the self-propelling power of the network is low. Dynamics means communication and information flow of network. If nodes x and y are not connected, the distance dist is zero (dist = 0). Further, the inverse length l^{-1} ranges from no edges to fully connected (0–1). In the context of botnets, l^{-1} introduces the covered circuit of bot-to-bot links fabricated by the malicious software, rather than the material anatomy of the internet. Therefore, victim bots of the same local network may have more than one hop apart or are not even connected in the bot network build by the malicious software. This metric function is pertinent because if all messages communicated through a botnet, there is a high risk of botnet failure. The precision required reveal probability of detection is not so easy to express, as botnet recognition is a new, evolving field.

Suppose bots x and y are linked via n likely routes, $R_1, \ldots, R_n$, and all nodes in the route can be revived (cleaned) with likelihood α. If ϵ_i is the possibility that route R_i is falsified, closed, then all routes between x and y can be closed with likelihood:

$$\prod_{i=1}^{n} \epsilon_i \leq (1-\alpha)^n$$

And nodes x and y are linked through path with probability $1-(1-\alpha)^n$, the failure likeliness rises with α. Under a hypothetical condition $l^{-1} = 1$, each bot can directly communicate to every other bot in the network. A botnet with many interlinks has more short paths, so it forwards commands speedily, and results in fewer detection possibilities.

To define the robustness of networks, a redundancy metric should be defined using local transitivity. Local transitivity reveals the possibility that nodes can be viewed in group of triplets. That means if there are two node pairs, {x, y} and {y, z}, i.e. {x, y, z} shares a common node y, local transitivity calibrates the likelihood that other two nodes, y and z, also have common edge. A clustering coefficient γ, calculates the mean degree of local transitivity in a community of vertices around node y, Γ_y. If E_y denotes the number of edges in Γ_y, then γ_y is the clustering coefficient of node y. Where k_y denotes the number of vertices in Γ_y, then

$$\gamma_y = \frac{E_y}{C_2^{k_y}}$$

$$\gamma = \langle \gamma \rangle = \frac{1}{N} \sum_{y \in V} \gamma_y$$

The average clustering coefficient $\langle \gamma \rangle$ determines the number of group of triplets divided by the maximum number of possible group of triplets. Like l^{-1}, γ falls in the range [0, 1] with 1 for a completely connected graph or mesh and 0 for disconnected graph. Botmasters mainly consider those targets that have acceptable attributes like low security levels, high transmission rates, low supervising rates, remote locations and easy availability. Ideally bots should run on those machines which have enough available bandwidth to paralyze any available service on the internet, and also it keeps the malicious activities hidden. The distance among compromised machines plays an important role in a botnet spanning different autonomous systems and nations, which makes it troublesome for law enforcement systems that are depended on to supervise unusual traffic to subvert it to track the activities of either.

6.1.2 Botnet Life Cycle

Whenever a botmaster wishes to infect some other target device, the botnet should descend through specific stages [1,5] and [11], as shown in Figure 6.2.

Phase 1 (Initial Infection): A botnet infects a new internet-connected device, then injects some destructive code. This goes through in various ways such as a virus infection, for example, through contaminated email attachments, automatic downloads of malicious software from websites, and so on.

Phase 2 (Secondary Infection): In this stage, the affected host seeks binaries of malware in an acknowledged network database by running a program. Malware installed for a period of time in Phase 1 discovers the repository of target's binary or the bot itself (or the C&C server), and should include the machine's address as hard-coded IP which can be static or dynamic. These addresses can be concealed directly as an enumeration of hard-coded IP addresses (static IP) or via a list of static or dynamic domain names, bringing

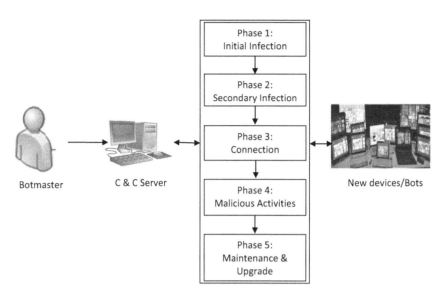

FIGURE 6.2
Botnet life cycle.

forth resistance to handicap the C&C server [13]. Whenever these binaries are downloaded and run by communication protocols like hypertext transfer protocol (HTTP), peer-to-peer (P2P) or file transfer protocol (FTP) protocol, and so on, then the host acts as a real bot.

Phase 3 (Connection): This phase is also well-known as a rallying mechanism, a particular action of connection set up with the C&C server. This stage is planned whenever the host is resumed, so that the botmaster is assured that the bot is alive and going to participate in the botnet formation and is in position to perceive commands to do unlawful tasks. Thus, this phase may happen multiple times during the entire lifetime of the bot. During this phase, bots are highly vulnerable because they have to contact C&C servers to get the commands. By default, this connection establishment creates some identifiable traffic patterns which may lead to identification of botnet components. The target automatically establishes a communication with the live C&C server.

Phase 4 (Malicious Activities): Through the C&C server, the bot army receives instructions from the botmaster to perform unlawful activities. During this phase, messages are exchanged more actively over a short time. Anomaly-based methods may or may not recognize botnet traffic because it is not in bulk and thus does not influence high degree of network latency. Malicious activities may include information theft, identity theft, stealing useful resources, monitoring network traffic, DDoS attacks, spreading malware, extortion, discover vulnerable computers, phishing, spamming, spoofing, and so on.

Phase 5 (Maintenance): This last step is generally used to update and keep the victims alive every time timely updates are sent to the zombie devices.

6.2 Infection Mechanism

There are different forms of methods to disseminate a specific bot: Email attachments, web downloads and automatic scans.

- Mail attachments: Email attachment is a download-based method. Email attachments with worms may come with bots. Spam delegates fast build-out of bots easily.
- Web download: Web-based malicious software builds structures similar to botnets, waits for commands and updates victim machines periodically to query web-based servers.
- Automatic scan: Automatic bot causes infection to the vulnerable host.

Botnet architecture: The C&C channel through which each bot forms a network may be categorized as per particular operational modes, topology and architecture [1]. The architecture followed by the bots to form a network are classified into three categories:

1. Centralized: This is the easiest to control and manage for the botmaster. The botmaster manages and oversees each and every bot in a botnet through a singular central point, the C&C server [5,10] and [11] as shown in Figure 6.3. Topologies used in this architecture are hierarchical and star topology. IRC and HTTP are commonly

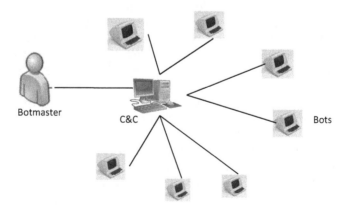

FIGURE 6.3
Centralized architecture.

used communication protocols. The strengths of this architecture are decreased regulatory cost, quick response time, harmony and easy observation of the stature of the botnet by the attacker, which provides several key pieces of information about certain primitive features, like the number of lively bots or their worldwide dispersion. The major difficulty of this architecture is the risk of failure is higher than with other architecture.

The protocols mainly used are Hypertext Transfer Protocol (HTTP) and Internet Relay Chat (IRC). In the IRC botnets, an IRC channel is created by botmaster on the C&C server through which the victim machines will communicate with each other to execute some unlawful and undesired activities. Despite the fact that IRC protocol is very flexible and mostly suited for C&C channel, it has some severe disadvantages because of the ease of detection and interruption of an IRC botnet operation. As IRC traffic is not usual traffic, its detection is easy and is seldom useful in corporate networks; actually, it is commonly shut-off. An admin may obstruct a botnet venture merely by noticing the network traffic and barricade it with firewalls. Because of limitations in IRC traffic, the leftover protocol is Hypertext Transfer Protocol (HTTP) which turns favorable. The positive aspect of HTTP is that this type of traffic is allowed in almost all networks, so the camouflaging the in-between communication of bots and botmaster is easy.

2. Decentralized: In this, no single entity is accountable for handling the bots in a network as shown in Figure 6.4. More than one C&C server is responsible for communication between bots. The detection of such a botnet is harder relative to the centralized architecture. There is no fixed C&C server so any compromised host can act as a server or as a client [5,10,11]. The design of this architecture is more complicated, and detection is also harder. Botnets which follows this architecture are not easy to stop because finding some or more bots definitely does not disrupt the whole network as there is no single controlling C&C server to be recovered and blocked.

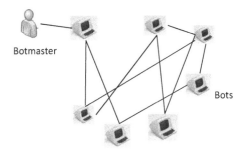

FIGURE 6.4
Decentralized architecture.

A decentralized botnet can be classified as per the different types of P2P protocols used, which are as follows:

- Structured P2P overlays: In this type of network, content and its location are mapped. For this type of network, routing is done through a distributed hash table (DHT). Example: Kademlia.

- Super-peer overlays: In super-peer networks, hosts are not equal. To reinforce network functions like control and search, a subset of the hosts is automatically picked out to be short-term servers. Some P2P applications are FastTrack, Skype, etc. Most efficient and effective botnets are not probably adopting this proto-type because these networks are more obvious and exposed to intended attacks. Bots that follow this basically have an existing IP address and are not listed under DHCP or firewalls.

- Unstructured P2P overlays: This includes arbitrary and unstructured topologies with different levels and state of dispersal or uniform arbitrary networks. They give no prospect for key lookup or routing. They support arbitrary walking, flooding and variations of the preceding as seeking procedures. The supported protocol is the gossip protocol.

3. Hybrid architecture: This is an amalgamation of both the types of architecture, centralized and decentralized, as shown in Figure 6.5. There are two types of bots in hybrid architecture: Servant bots and client bots. Servant bots acts as both client and server [5,10,11]. They are tacked together with routable and fixed IP addresses, whereas client bots not configured to welcome incoming communications and with routable IP addresses. They are positioned behind fire-walls without internet connections. Servant bot IP addresses are listed on the predefined peer lists. They attend to a specified port for incoming communications and for this communication, they utilize an autogenic symmetric key encryption, which causes extra diffi-culty in botnet detection. Each and every bot compulsorily makes

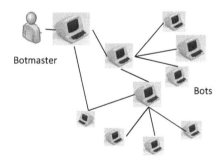

FIGURE 6.5
Hybrid architecture.

connection with servant bots which are listed on their respective peer list at a fixed interval to fetch commands given by their botmaster. Whenever a bot receives instructions, it quickly instructs every servant bot on their defined peer list. Detection of this type of botnet is more difficult than other types of botnets.

6.3 Botnet Evasion Techniques

Bots have become progressively ultra-modern, so their evasion methods have been flourishing to avoid detection mechanisms so that botnets can be operated for a long time. Botnet fitness can be linked in terms of its functionality. Its usefulness is reduced if a botnet's evasion mechanism doesn't work properly. So, as per the degree of complexity, the evasion technique thus needs to be enhanced and a new one should also be developed.

Various different methods are used, including fast-flux service networks (FFSNs), tunneling through HTTP, ICMP, statistical patterns changes, encrypted traffic, arbitrary communication patterns of bots, different task bots in the same network and using dynamic DNS entries [3]. Initially detection techniques were looking for unusual communication between bot to bot and bot to botmaster or vice versa. To conquer this, bots advanced and practiced a cipher algorithm. Thus, payload inspection is not so powerful [6]. However, detection based on clustering schemes is convincing, because of arbitrariness in communication patterns and designating different responsibilities to each and every bot. Various studies shows that bot make use of random high-numbered ports. Each bot picks out self-generated random high-numbered ports which it listens in on that lies between 1,025 and 65,535, and mostly avoid all detection cases.

Fast-flux service networks (FFSNs): The lattermost botnets use FFSNs as their command and control (C&C) mechanism. It is a Domain Name Service (DNS) technique in which compromised hosts change their network to hide phishing. To establish anonymous communication over networks, bots first make connections to a victim, who acts as a proxy, progressively forwards commands from the bot to the C&C server and redirects reactions to bots from the C&C server. FFSNs amalgamate round-robin IP addresses and short Time To Live (TTL) values for some known specific DNS Resource Record (RR) to allocate instructions to several victims. Both services HTTP and DNS are always hosted in all nodes in FFSN to concomitantly control accessibility of content for many of domains on an individual host [7]. Another silent approach for tracking and detecting malicious FFSNs is passively examining recursive DNS (RDNS) traffic data gathered from several large networks.

Domain flux: This is a technology in which a harmful botnet remains hidden by repeatedly changing the domain name of C& C server of the botnet. Several domain names are generated using an algorithm to employ this technique which is known as domain generation algorithm (DGA) [8]. DGA is executed by attackers to dynamically create pseudorandom domain names. It has multifaceted extraordinary features. Foremost, it is not necessary to hard-code the C&C domains in the binary file of the malware. Second, DGA can also be used for safe pull-back scenarios whenever the initial communication fails; it means botnets can recover easily from failure. DGA can be implemented in various ways. First, by creating a hash value and converting it in an ASCII string and appending it with the top-level domain (TLD), such as .com or .org, and so on. In order to produce many domains, the logic should be employed in a repeated fashion. Generation of many unique DNS names can be done in this way and then one can be chosen and used for C&C communication.

6.3.1 Botnet Defense

A botnet is a different version of an attack which has originated from the conventional malicious code, so the defenses of botnets still relate to the strategies used to defend against malicious code. The first and foremost step in defense against botnets is the prevention approaches used so that system security is defined at the start, but if some illegal activities are already going on, then detection techniques are used and responses against attack are prepared to deal with losses and the safety of remaining resources. Whenever a botnet is discovered, the very first remedy is to either halt the bot or discontinue the entire botnet. Blocking some bots is not efficient for unraveling the issue, as there are many more victims in the network. More productive methods are akin to paralyzing the entire network.

Techniques of defending against bots pivot around two key actions [5]:

1. Propagation of bots: Battling bot/worm attacks immediately impacts the number of victim hosts in a network, thus lessening the throttle-hold of the network and hence its usefulness to the botmaster.

2. Bot communications: Another type of defensive action is to cease the connection between bot-to-bot and bot-C&C servers and vice versa by unhinging the communication, so that the attacker is not able to convey instructions or carry out malicious activities.

The heart of the above defense approaches revolves around three key areas: prevention, treatment and containment.

6.3.2 Prevention

Prevention confines the influence of Botnet attacks, such as by isolating the compromised machines. Basically, three mechanisms are used: endpoint security, intrusion prevention systems (IPS) and vulnerability management.

Vulnerability management is a recurrent routine of discovering, classifying and lessening vulnerabilities. Insider threats are the biggest threats. Security and management of endpoints and security policies of network devices should be enhanced.

For effective defense, some actions should be carried out by network experts, administrators and Internet Service Providers (ISPs). Key features malware dispersion must hold are the number of unprotected hosts, infection continuance and infection flux. Thus, the aim of these prevention techniques is to reduce the number of vulnerable machines, the level of malware outspread, and the size of the botnet. Preventive actions include system maintenance: Secure software development, use of antivirus program, vulnerability removal, and so on. There is no assurance that software is fully secure which means no software is perfect and users must have adequate understanding of this so that one cannot claim usage of network devices is riskless.

6.3.3 Treatment

Treatment relates to autoclaving zombies to lower the bot count and to lessen the number of unprotected hosts and the malware outspread rate.

Containment: The containment stage is an amalgamation of two stages; one is botnet detection and the other is response.

Botnet detection: Botnet detection is very crucial task to enhance the internet security. It is usually done by tracking live hosts and/or network [5].

There are several data types used in detection techniques [1]. These are DNS data, packet capture data, netflow data, host data and honeypot data. Organizations analyze networks, systems and transaction logs. DNS is a protocol which acknowledges each and every query with a respective resource record (RR). The dynamic structure of the DNS makes it captive for botnets to misuse the system for performing various malicious tasks. A flow can be imagined as sending IP packets in and out from one point to another in the network during a pre-specified distance and time. Honeypot data enables officials to perceive botnet activity. According to the various theories and conducted experiments, detection techniques can be categorized into honeynets detection techniques and intrusion detection techniques [4,9].

Honeypots and honeynets both denote that the end devices like PCs are the prime modus operandi to gather crucial data about the attacks. Because these devices are quite vulnerable to malicious attacks, they are quite easy for botmaster to attack and compromise. Honeynets are important for perceiving the botnets' properties because botnets change their behavior and impact regularly to thwart their detection.

IDS (Intrusion Detection System): IDS is useful in supervising the netflow for the abnormal actions going into a network. It directly notifies the administrator of the system and takes action against it if it notices some malicious activities during traffic. Two types of IDS are anomaly-based and signature based.

1. Anomaly-based detection [9]: Anomaly based technique accepts only those network activities which are clearly pre-stated by the network administrator. In advance a set of rules should be pre-stated and each rule should be examined for its correctness. It identifies actions which does not follows stated rules. From a computational viewpoint this technique is a little bit costly but more reliable than signature based. Some disadvantages are also associated with this technique that the definition of rules is not so easy, different rules are defined for different protocols. This technique comes with some limitations like more time required for supervising the bot infected hosts. This detection technique is also divided into network and host-based techniques.

2. Signature-based detection [9]: The most edgy part of this technique is that activity patterns or signatures are so easy to perceive if performance of network is already known. This technique is too simple to interpret and develop. Nowadays, botmaster employs timely change in attack pattern to launch an unshakable attack.

6.3.4 Response

This stage relates to methods to cease the communication between bot-to-bot and bot-to-C&C servers and vice-versa and, finally disable the server. This can be practiced using automatic methods that incorporate content filters, firewalls, IP-address blacklisting and cease communications in-between bots and malware overspread to minimize or halt the distribution of bot commands, discontinue the botnet connection and finally disable the C&C servers.

There are two common responses against botnets: null routing and quarantining [12]. The quarantine is a state of enforced isolation to segregate and limit the mobility of computer. A null route is a passage to real no-where in the network. As per this routing packets fall down instead of send. The behavior of null routes is usually known as blackhole filtering. Null routes are normally implemented with a unique flag or remark set to route, but also can be configured by sending packets to a prohibited IP address such as loopback address or 0.0.0.0. This technique has supremacy over traditional firewalls as its availability on each and every possible network router and puts essentially no performance issue. Null routing can always experience higher throughput than classic firewalls because of higher-bandwidth routers.

6.3.5 Source

Botnet detection sources can be classified into virtual network, real network and honeypot traffic. Virtual networks are superintended networks on which botnets are manually executed to learn the botnet behavior. Real network monitoring is a white collar answer to keep the continuous record of network traffic with all provided data and learned facts. A honeypot is a purposely created ground to explore the attack pattern and their lethal motives.

6.3.6 Algorithms

There are various algorithms for botnet detection and mitigation which can be categorized into pattern recognition, clustering, correlation, heuristic rules, fuzzy and statistics. In decision trees, classification rules are represented by the routes from root to leaf. A clustering algorithm presents an effective metric for learning intrinsic models of network traffic without the required labels to instruct the algorithm. The significance of the heuristic algorithm is that it doesn't require prior knowledge about system behavior for drawing out solutions from multi-facets. In fuzzy logic, a set of rules are derived from known specifications of the botnet. Correlation enhances sensitiveness and a clear picture of problem. Correlation flourishes certain patterns to perceive the well-known and detectable patterns and develops unspecified patterns for detecting the unrevealed and the unnoticeable activity. A flow-based system believes in supervised learning for differentiating botnet traffic. Machine learning gains knowledge from a computational learning model and pattern recognition and analysis of algorithms that can acquire knowledge and make projections on provided information.

6.4 Botnet Applications with WoT

A botnet is a set of gadgets like computers, cellular devices, servers, and so on. which might be connected to the net. Each of these gadgets is hosting one or greater bots; these are infected and controlled by a commonplace type of malware. Botnets can be used to carry out disbursed denial of provider assault, steal records, ship unsolicited mail and allow the attacker get right of entry to the tool and its connections. Users are usually ignorant of the botnet infecting the device. As botnets as self-reliant, they are not choosy of what devices to contaminate, hence they infect anything with a web connection. With the spread of the Web of Things, gadgets are increasingly entering the pool of capability candidates for a botnet. Even worse, with the Web of Things nevertheless in its teething stage, security for most of these gadgets is vulnerable.

- Mirai Botnet: In 2017, the Mirai botnet attacks commenced. It scanned the web for WoT devices, then tried 60 default usernames and passwords to get the right of entry to them. Once successful, the assault inflamed the compromised device with the Mirai botnet malware. With its swiftly forming army, Mirai started out to attack websites across the internet. It did this by using the use of its army to perform DDoS attacks, swarming websites with connections from the gadgets on the botnet.

- Torii Botnet: In 2018, a brand-new botnet assault came into life: Torii. Unlike the other WoT botnets that used Mirai's code, this one became its own strain. It used fantastically superior code, capable of infecting a huge majority of net-connected gadgets. Torii hasn't attacked something just yet; however, it is able to collect an army for a large assault.

- Mad WoT: An examination by way of Princeton has proven that WoT botnets may additionally hold the strength to take out electricity grids. The record describes a method of assault known as "Manipulation of demand via WoT" (MadWoT), which acts like a DDoS attack but its objective instead is the power grid. Hackers could deploy botnets on high-energy WoT devices, and then authorize all of them at the same time to cause a blackout.

6.4.1 Botnet Size Measurement

Botnet size is the key metric amongst various metrics used to measure the seriousness of the botnet threat [14]. Former studies of this metric generally revolve around the details and existence instead of the essence of the problem, for example, versatility, complexity of botnet behavior. Quantitative analysis of botnet size should be done with the following metrics:

1. Botnet live population measurement: From the viewpoint of evaluation of threat seriousness, the botnet live population denotes the magnitude of attack and can be realized by anomaly detection. This can be done as follows: Active/passive DNS detection – active DNS detection depends on how progressively the domain name of C2C server is utilized. DNS redirection monitors communication between various bots and C2 server. The active volume of the botnet can be determined by enumerating the number of hosts who are involved in the communication. Passive DNS detection denotes specific pattern DNS query collected passively over the network. General C2C communication features include: stable pattern of botnet C2C communication, URL in spam, unusual in/out degree of hosts, unusual flow in network, and so on.

2. Botnet footprint measurements: For in-depth sight of the botnet, its footprints must be evaluated. The efficient and accurate method is to first decide whether host-based misuse or anomaly-based infection

is in use for detection, and then count the victims. It can be determined by statistical inference.

3. Dynamic botnet size: Botnet detection may show results differently in different time zones and locations so it plays an important role in botnet size measurement.

Tracing dynamic botnet size incorporates some of the following facets:

- Patterns of botnet propagation – Botnet propagation depends on how effectively it scans vulnerability. Scanning patterns can be of worm-type and non-worm class. Worm class is the basic manner in which the botnet has a huge volume to scan and has large number of infected machines in a short time period. The non-worm class has wide varied scanning algorithms, which include network scanning, hit-list and random scanning.
- Obscure pictures generated by some activities of botnets, such as botnet cloning and migration, which are used to examine the ownership of groups of bots.

6.5 Conclusion and Future Work

In this chapter, the topic of botnets including their architectures, communication protocols, detection techniques and algorithms are discussed. As today's botnet comes with various evasion techniques and tricks to make fools of network administrators, a part of future work may include case studies and the various hands-on tactics to deal with the new versions of botnets, such as analysis of DNS queries, studying various cryptographic techniques which are used to hide the network, and so on.

References

1. Sérgio S.C. Silva, Rodrigo M.P. Silva, Raquel C.G. Pinto, Ronaldo M. Salles. (2012, October). Botnet: A survey. *Computer Networks*, 57 (pp. 378–403).
2. David Dagon, Guofei Gu, Cliff Zou, Julian Grizzard, Sanjeev Dwivedi, Wenke Lee, Richard Lipton. 2005. A taxonomy of botnets. In: *Proceedings of CAIDA DNS-OARC Workshop*.
3. Matija Stevanovic, Jens Myrup Pedersen. (2016, January). On the use of machine learning for identifying botnet network traffic. *Journal of Cyber Security and Mobility*, 4(2) (pp. 1–32).

4. Gao Jian. (2017, February). Review of the research on botnet. *International Journal of Computer and Applications* 160(3) (pp. 0975–8887).
5. M. Stevanovic, J.M. Pedersen. (2013). Machine learning for identifying botnet network traffic.
6. Xingguo Li, Junfeng Wang, Xiaosong Zhang. (2017, September). Botnet detection technology based on DNS. *Future Internet* (pp. 55–77).
7. Sheharbano Khattak, Naurin Rasheed Ramay, Khan Kamran Riaz, Affan A. Syed, Syed Ali Khayam. (2014). A taxonomy of botnet behaviour, detection and defense. *IEEE Communications Surveys and Tutorials*, 16(2).
8. Zou Futai, Zhang Siyu, Rao Weixiong. (2013, November). Hybrid detection tracking of fast flux botnet on domain name system traffic. *China Communications*.
9. Abdullah Raihana Syahirah, Mohd Faizal Abdollah, Zul Azri Muhamad Noh, Mohd. Zaki Mas'ud, Siti Rahayu Selamat, Robiah Yusof. (2013, March). Revealing the criterion on botnet detection technique. *IJCSI International Journal of Computer Science Issues*, 10(2, No 3) (pp. 208–215).
10. Wazir Zada Khan, Muhammad Khurram Khan, Fahad T. Bin Muhaya, Mohammed Y. Aalsalem, Han-Chieh Chao. (2015). A comprehensive study of email spam botnet detection. *IEEE Communication Surveys and Tutorials*, 17(4) (pp. 2271–2295).
11. Abderrahmen Mtibaa, Khaled A. Harras, Hussein Alnuweiri. (2015, August). From botnets to mobibots: A novel malicious communication paradigm for mobile botnets. *IEEE Communications Magazine* (pp. 61–67).
12. Elisa Bertino, Nayeem Islam. (2017, February). Botnets and Web of Things security. *IEEE Society* (pp. 76–79).
13. Aditya K. Sood, Sherali Zeadally, Rohit Bansal. (2017, July). Cybercrime at a scale: A practical study of deployments of HTTP-based botnet command and control panels. *IEEE Communications Magazine* (pp. 22–28).
14. Liu Shangdong, Jian Gong, Wang Yang, Ahmad Jakalan. (2011). A survey of botnet size measurement. *Second International Conference on Networking and Distributed*.

7

WoT-Enabled Smart Cities

Aarti Jain and Rupali Rani

CONTENTS

7.1 Introduction

With the spread of urbanization, cities in today's world require intelligent solutions to tackle critical issues like healthcare, energy management, transportation and infrastructure. The Internet of Things (IoT) is the most likely technology for tackling these challenges through automation, networking, sensing and proper data analysis. Smart cities are the concept of utilizing new technologies and connected data sensors using wireless or wired communication to enhance and become powerful in term of infrastructure and city operations. WoT (the Web of Things) is a system that comprises thousands or more gadgets and sensors which communicate with each other. The artificial intelligence, R programming, Python and machine language are capable of helping the network to process the information received from the connected gadgets. These technologies are also supported for adjusting, monitoring and managing devices. Basically, a smart city is the urban area surrounded by or embedded in a smart system. The major use cases of a smart city are smart transportation systems, smart parking, smart building monitoring, smart agriculture, smart waste management and smart security systems.

7.1.1 Definition

The International Telecommunications Union (ITU) (ITU, 2014) emphasizes information and communication technologies (ICT) and considers a smart sustainable city as

> An innovative city that uses information and communication technologies (ICTs) and other means to improve quality of life, efficiency of urban operation and services, and competitiveness, while ensuring that it meets the needs of present and future generations with respect to economic, social and environmental aspects.

Similarly, the International Standards Organization (ISO) (ISO, 2014b) recognizes the smart city as

> A new concept and a new model, which applies the new generation of information technologies, such as the internet of things, cloud computing, big data and space/geographical information integration, to facilitate the planning, construction, management and smart services of cities.

Moreover, it defines smart city objective to pursue: Convenience of the public services; delicacy of city management; livability of living environment; smartness of infrastructures; and long-term effectiveness of network security.

Furthermore, the British Standards Institute (BSI, 2014) defines a smart city as

> The effective integration of physical, digital and human systems in the built environment to deliver a sustainable, prosperous and inclusive future for its citizens.

7.2 Components of Smart Cities

Smart cities are responsive, intelligent, connected and sustainable. The vital segments of territory-based advancement in the Smart Cities Mission are city improvement (retrofitting), city reestablishment (redevelopment) and city augmentation (greenfield advancement) in addition to a pan-city activity in which smart solutions are applied, covering bigger pieces of the city. The main components of smart cities are:

Smart mobility: Smart mobility refers to intelligent traffic planning, smart interconnection of all roads and a smart parking management system.

Smart healthcare: Smart healthcare refers to use of digital and mobile technology for advancement of eHealth and mHealth system for the monitoring of patient health.

Smart governance: The real measure of a smart city is people's happiness. Smart governance means we are constructing a city where all the resources are efficiently utilized. We are protecting both our people and our information; providing a smart economy for business visionaries and entrepreneurs; an enhanced quality of life for people, with simple access to customized city administrations, protecting and nurturing our natural environment (Figure 7.1).

Smart security: Smart security refers to the connected devices that should be protected by an advanced IoT security solutions.

Smart environment: This refers to innovation in natural resource protection and management, recycling of waste products, pollution control, planning of green areas and green energy.

Smart buildings: Smart buildings make the use of real-time control of building solutions like heating, ventilation air conditioning, smart lighting and security, etc. for enhancing the quality of life of the people.

7.3 Iterative Approach to Implementing a Smart City

The scope of smart city applications is exceptionally varied. What they share, for all intents and purposes, is the way to deal with usage. Regardless of the work performed, they should begin with the establishment – a fundamental

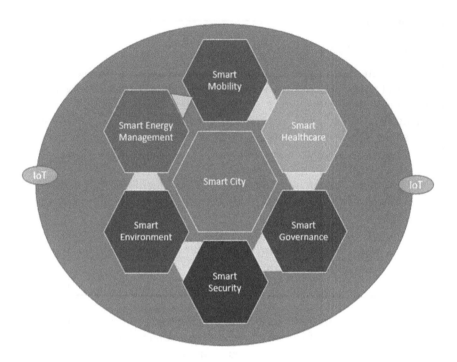

FIGURE 7.1
Components of smart cities.

keen city stage. In the event that a region wants to grow the scope of smart city benefits in future, it will be conceivable to overhaul the current engineering with new devices and innovations without remaking it. There are six step implementation models to pursue for making a proficient and adaptable IoT design for a smart city.

7.3.1 Implementing a Smart City Platform

To have the option to scale, smart city execution should begin with planning essential engineering – it will fill in as a springboard for future upgrades and permit including new administrations without losing practical execution. The IoT solution for smart city includes four components which are network, gateways, data lake and data warehouse.

1. *Network*: A smart city – just as any IoT framework – utilizes smart things furnished with sensors and actuators. The main objective of sensors is to gather information and pass it to central cloud management system. Actuators enable gadgets to act, such as by modifying the lights colors, confining the progression of water to pipes with leaks, and so forth.

2. *Gateway*: Any IoT framework involves two sections – a "tangible" portion of IoT gadgets and system hubs, and a cloud part. The information can't just go from one section to the next. There must be entryways – field gateways. Field gateways gather the data, compress it and filter information before moving it to the cloud. The cloud platform guarantees secure information transmission between field gateway and the cloud portion of a smart city arrangement.

3. *Data lake*: The principal work of a data lake is to store information. The data lake safeguards information in its crude state. At the point when the information is required for important bits of knowledge, it's extricated and passed to the data warehouse.

4. *Big data warehouse*: A big data warehouse is a repository which contains structured data. After defining the data value, the data is extracted, modified and loaded into the big data warehouse (Figure 7.2).

7.3.2 Monitoring and Basic Analytics

The data analytics help to monitor and analyze device environment. It also sets rules for device control. For example, to measure the soil moisture level, the sensors deployed are used to set rules for control of valves for different moisture levels. The data collected can be viewed on different cloud platforms.

7.3.3 Deep Analytics

To process IoT generated data, the hidden correlations between sensor data are monitored, analyzed and identified. Advance techniques like machine learning (ML) algorithms and statistical analysis are used for deep analytics. The ML algorithm predicts the future value of the data stored in the data warehouse based on the past experience. The models are utilized by control applications that send directions to the IoT gadget actuators.

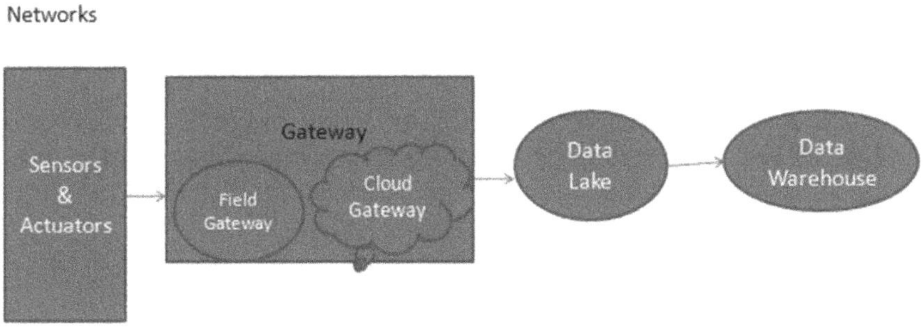

FIGURE 7.2
IoT component solutions for smart cities.

7.3.4 Smart Control

The commands are sent to actuators to ensure better automation by sending command signals. The command signal directs actuators what to do to perform particular tasks. There are two types of command signals which guide the actuators. One is rule-based and another is ML algorithm-based. The rule-based command signals are defined manually while the ML-based control applications use models made by ML calculations. The models are recognized based on data analytics. They are tested, approved and regularly updated based upon requirement.

7.3.5 Interacting with Citizens via Client Applications

Along with control applications, the WoT-enabled device must have the privileges of interacting with the user. It should have a central cloud platform where the users can perceive the data, and monitor and control the device. The client application allows users to connect with central smart city management. For example: GPS enables the user to see the traffic jam on a smartphone. It also provides notification to follow a different route to avoid congestion.

7.3.6 Integrating Several Solutions

To become smarter is a continuous and ever-ongoing process. With the help of ideas and modern technology, all the solutions are integrated together continuously and steadily to increase the function in a single device. For example: Some years ago, we had a traffic management system which provided traffic congestion information. With the help of sensors, we implemented the pollution monitoring system and traffic congestion system in a single traffic management system, as shown in Figure 7.3. The iterative approach and integration of solutions helps to reduce cost, provide faster payoff and provide information on a single platform.

7.4 Major Use Case of a Smart City

The major use case of smart city includes smart infrastructure, smart buildings, smart home security systems, smart waste management, smart environment and smart healthcare.

7.4.1 IoT-Enabled Smart Infrastructure

Smart infrastructure includes smart lighting management, smart parking management system, smart transportation facilities, connected streets and a charging system for vehicles.

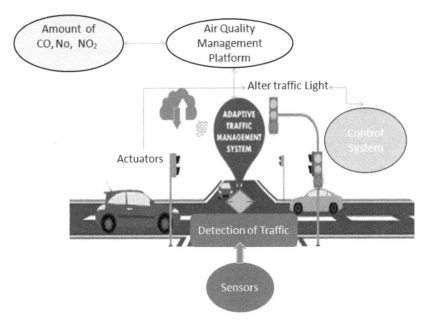

FIGURE 7.3
Integration of traffic management and air quality management.

1. *Smart lighting*: With smart lighting, a demand-based lighting system can be provided. Users can change the color or brightness according to requirements and even control them from a smartphone anywhere in the world. It helps in daylight harvesting and improves energy efficiency. LED technology is used in smart city initiatives.

2. *Interconnected streets*: Interconnected and smart streets obtain data and provide information and services from millions of sensing devices. It provides information about locally detectable danger, traffic condition, road blockages and roadway construction, etc. The system driver is made aware of the condition of roadways and thus able to save time and energy.

3. *Smart parking management*: Smart parking management systems provide information about the available parking space in real-time scenarios. Different types of sensors like infrared, passive infrared and ultrasonic sensors are used to find the unoccupied location of vehicle at public place. These sensors are embedded into parking spaces, transmit data at regular times and provide information about the parking space via digital signal processors into a central parking management application. Smart parking reduces traffic congestion; saves time and resources; decreases harmful gas emissions from vehicles; lowers management costs and driver stress.

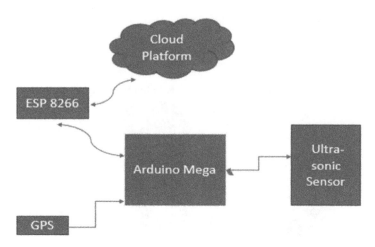

FIGURE 7.4
Smart parking management system.

For powerful deployment of smart stopping innovations, every gadget needs to have a dependable availability with the cloud servers.

- A microcontroller Arduino Mega 2560 is integrated and interfaces with other components. The components include a GPS module which is used to navigate the parking slots for the driving users. Ultrasonic sensors are placed in the parking space that sense the presence of vehicles and accordingly transmits the information to the user. ESP 8266 is used to send the information to a cloud platform (Figure 7.4).

4. *Connected charging stations*: As a battery is the main device for energy storage in electric vehicles, it needs to be charged when required. The smart infrastructure for the battery includes connected charging stations for the electric vehicles at different places like parking area, shopping malls, city fleets, airport etc. The vehicle charging platform can be integrated with IoT to display the battery status. The user can view the data and locate nearby charging stations using the app.

5. *Intelligent transportation systems*: These aim to provide innovative, efficient and reliable services related to different modes of transportation and traffic management. The main services of ITS are

- Bus information services (BIS): It provides information about the arrivals, departures and routes of the buses.
- Electronic toll collection system (ETCS): An electronic toll collection system is an automatic system to collect tolls from various methods of payment like coins, smart cards, tokens and

credit and debit cards without the need for a toll collector. These administrations plan to dispose of the delays on toll streets, carpool lanes, toll extensions and toll tunnels. The system uses an E-Pass which has a transponder mounted on a vehicle. The information about the vehicle and user is stored in the transponder. The antenna at the toll booth senses the transponder and deducts the toll charge, which allows the vehicle to pass.

- Automatic fare collection (AFC): AFC helps in automatic collection of fares. It makes use of smart cards, transponders, and automatic gate machines and ticket vending machines for the collection of fares.

- Automotive navigation system (ANS): It is used with satellite navigation systems to monitor the position of automobiles on roads. Dead reckoning takes distance data from sensors attached to the drivetrain, a gyroscope and an accelerometer. Automotive navigation is based on the shortest path concept of graph theory.

- Eco-driving services: Eco-drive support services provide information about driver's fuel-efficient driving. By using an eco-drive support service, individual drivers can reduce CO_2 emissions from their vehicles.

7.4.2 IoT-Enabled Smart Buildings

Smart buildings have the ability to analyze the environment and make real-time adjustments to improve efficiency and productivity. Smart buildings integrate and collect information from different embedded devices or sources for intelligent control of smart building devices. A smart building has different infrastructural components that maintain the occupant's comfort level and provide quality of life to people. Some of them include highly efficient HVAC systems, smart metering systems (electricity, gas, water), occupancy monitoring systems and vehicle charging technology. Various systems are used to maintain the complete health of the surroundings by monitoring all the assets and ensuring the safety and security of buildings. The best example is the Alexa-based home automation-controlled building.

7.4.2.1 Challenges for Smart Buildings

- Low resource usage
- Security integration
- Increased occupant comfort against varying temperatures
- Proper environment monitoring
- Proper network connectivity

7.4.2.2 Solutions for Smart Buildings

- *Analysis of advanced building*: Continuous analysis of sensor data from building systems helps in improving automation rules and hence reduces energy and water consumption.
- *Implementation of safety and security systems*: Various monitoring systems are implemented like IP surveillance, biometrics, wireless alarms, CCTV cameras, etc., to reduce unauthorized access and chances of thefts in a building, thus guaranteeing safety and security.
- *Application of machine learning algorithms*: Algorithms are built so that the environment factor of all building zones is captured to maintain a constant temperature throughout.

7.4.2.3 Case Study for Smart Buildings

7.4.2.3.1 Intel Smart Building Solution

A smart structure arrangement dependent on its IoT reference engineering is planned by Intel with the envisioned plan of expanding vitality preservation, utilitarian proficiency, individuals' comfort and security conditions. IoT innovation gathers the information from different structure framework like ventilation, cooling (HVAC), water, vitality age, tenant counters and the electrical framework. It further breaks down the information, sends it and stores it on the cloud web. The structures additionally have an ethernet-based brilliant lighting framework with sensors fixed in brightening devices to monitor temperature, occupancy and other natural variables.

The IoT gateways based on Intel processors control the entire system of the buildings. It uses different types of protocols such as Modbus TCP/IP and BACnet-IP to send and collect messages from the entire building system. Different varieties of smart sensors are associated safely to screen the structure's frameworks. Sensors also guarantee an uninterrupted flow of data between different devices. IoT gateways help in data integration and provide information to the entire solution. To provide end-to-end security protection of network and data, the gateway is protected by enterprise-grade security features, such as McAfee Embedded Control (Figure 7.5).

The building management systems are integrated and analyzed via iBMS. iBMS software runs on a different computing environment integrated with gateway software components. The gateway process message based on some predefined rules. The rules are divided between the gateway and the back-end server. When the data is calculated on the gateway, it is further transferred to the cloud platform over the MQTT communication protocol. The data is generated by sensors and is distributed across IoT gateways. To protect data from intruders, various network topologies and security policies are designed via the sensor network. The firewall is used to protect

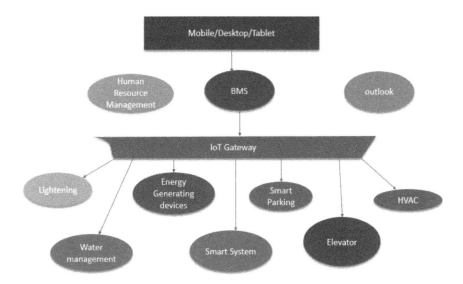

FIGURE 7.5
IoT-based smart building management.

the software running on the server and no internet connection outside the firewall is permitted. Data transmission security is guaranteed by open SSL (secure socket layer v2/v3) and transport layer security (TSL V1). The gateway bears a trusted platform module chip to scramble the application and security conveyance. The gateway security is ensured by means of secure boot and application whitelisting software. Secure boot checks the gateway working framework condition and application whitelisting guarantee that only specific set of application are allowed to run on the gateway (Figure 7.6).

7.4.2.3.2 Google Cloud Platform (GCP) Technology Used in Smart Building Platforms

- *Cloud IoT*: This is a fully managed service to connect the system to the cloud platform. It also collects, processes and analyzes data.
- *Cloud pub/sub*: This is an enterprise message-oriented middleware that provides low-latency, durable information that helps designers rapidly incorporate frameworks facilitated on the Google Cloud Platform and remotely. It has a high bandwidth. API is an example of cloud pub/sub.
- *Cloud data flow*: This provides unified batch and streaming process services. It is an open source programming model and is fully managed.
- *Big table spanner*: This is a fully managed service for data storage. The big table spanner is scalable and no SQL database is required. A data lake is an example of big table.

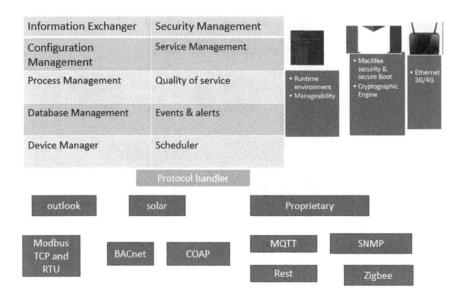

FIGURE 7.6
Gateway stack of Intel smart building solutions.

- *Building ontology*: A building ontology engineer collects the information from the building expert and accordingly makes the model called a modelet. After they develop the ontology engine algorithm about the building architecture, like single stories, multi-stories, bathroom area, kitchen area, games area, window area etc. to provide an ambient environment to the buildings (Figure 7.7).

7.4.3 IoT-Enabled Home Automation Security System

The IoT is a new innovation which is widely used for the development of smart home systems in order to provide intelligent safety, comfort and improved quality of life. Smart home automation is a user-friendly service for smartphone users. The technology uses a low-cost wireless controlled smart home system for controlling and monitoring the home environment. It includes remote monitoring, IP surveillance camera, biometrics and wireless alarms to reduce unauthorized access. A micro web server with IP-based connectivity is used for controlling and accessing devices remotely from an Android app. Raspberry Pi is used as micro web server which requires user authentication in order to access home automation system. A single sensor or multiple sensors are used with a single device to collect data from sensor nodes. A collection of sensor nodes within a distributed environment is called a sensor network. A data communication link is established to transmit data to a central data location node.

Home automation security control includes motion detection and video monitoring to detect the presence of authorized or unauthorized people.

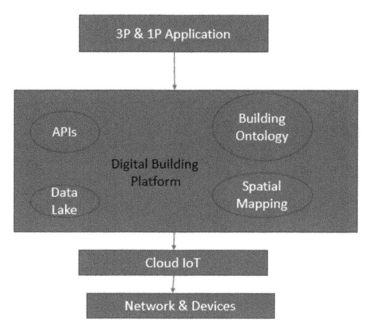

FIGURE 7.7
GCP technology used in smart buildings.

Security cameras are also used for proper surveillance. Digital monitoring allows owners to perceive any area of home via webcam from anywhere with internet connectivity. The security cameras are backed up to the cloud server to review home activities at any time. If any intruders try to enter in the restricted area, the PIR sensors detects human body motions by measuring changes in the infrared levels emitted by surrounding objects. Whenever a motion is detected, the alarm buzzes and an SMS or email is sent to the concerned person (Figure 7.8).

7.4.4 IoT-Enabled Waste Management System

Garbage accumulation is one of the common problems of a big city. In order to make the environment neat and clean, necessary action should be taken to collect the garbage and dump it properly. IoT-based waste management is a smart waste management system in which devices are embedded to send data to a cloud platform when a dustbin in different locations reaches a threshold level. The device will transfer the information about the dustbin level to the closest authority via SMS or email. The AT command is used to provide a messaging service through a GSM module. The program is burnt into a microcontroller through the Arduino software. The cleaning authority collects the garbage from the defined location and updates the status.

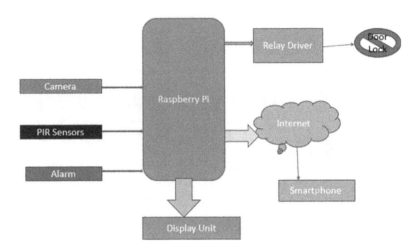

FIGURE 7.8
Home automation security system.

- **System Module**
  - *Ultrasonic sensor*: The ultrasonic sensor is utilized to recognize the level of the trash in the residue canister. It makes an ultrasonic pulse called a ping and looks for the reflection of the pulse. The sound pulse is made electronically, utilizing a sonar projector which is comprised of a signal generator, power amplifier and electro-acoustic transducer cluster. The ultrasonic sensor utilizes the data along with the time contrast between the sending and accepting of the sound pulses to decide the distance to an item.
  - *RF module*: The radio-frequency transmission (RF transmission) system is based on Amplitude Shift Keying (ASK) modulation technique using a transmitter/receiver (Tx/Rx) pair operating at the 434 MHz frequency band. The signals are taken as serial input by the transmitter and are transferred through RF. The receiver module then receives these signals. The system allows simple communication between two nodes which can be either transmission or reception at a single time. The encoder converts these parallel inputs into a serial set of signals and transfers them to the reception point through RF. These serial signals are further decoded using a decoder and the original signals are retrieved as outputs. These outputs can be observed on corresponding LEDs.
  - *Arduino mega*: This is a microcontroller board based on the ATmega328. It consists of a 14-digital input/output pin of which

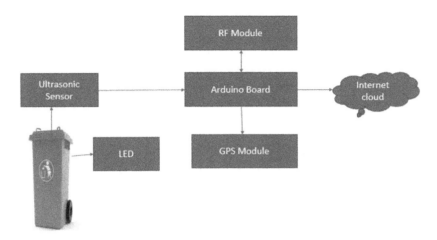

FIGURE 7.9
IoT-based waste management.

six pins are used as PWM outputs and six pins for analog input, a 16 MHz crystal oscillator, a USB connection, a power jack, an ISCP header and a rest button. It is connected to a computer by a USB cable or is powered with an AC to DC adapter or battery.

- *GPS module*: A GPS module is a global navigation system which provides the exact information about the dust bin on the cloud platform.
- *Internet cloud*: The internet cloud uses the cloud-based data collection technique and mobile app-based monitoring system for the dustbin level indication at different positions.
- **Advantages**
  - Real-time information about the dustbin level
  - Intelligent management of services
  - Resource optimization
  - Cost reduction
  - Environment quality improvement (Figure 7.9)

7.4.5 IoT-Enabled Healthcare

The main goal of a smart city is to provide a good quality life to its citizens and smart healthcare helps to accomplish the goal. According to a survey by Transparency Market Research, the smart healthcare product market is expected to reach by $57.85 billion by 2023. Smart healthcare is not for just curing disease, but also includes early detection and prevention of disease.

IoT and artificial intelligence play a key role in real-time patient health monitoring. IoT allows collection of data from smart devices and sensors to provide information about various parameters of body. The miniature sensor devices are mounted on the surface of the body or embedded inside the body. The sensors communicate with medical devices and measure various physiological parameters like blood pressure, blood flow, pulse rate, body temperature, and so on. The data collected are analyzed by remote servers. Artificial intelligence technology is applied to perform the task of analyzing the laboratory test like X-rays, CT scans and data entry. AI-based apps provide access to the current medical conditions of a patient which provides assistance in medical consultation from doctors. Technology like blockchain and smart cards are used to maintain patient electronic health records and link the patient to the payment gateway and insurance (Figure 7.10).

A smart drug acknowledgment gadget is likewise utilized on Android-based cell phone applications to scan the prescription and in this manner helping persistent opportune taking of medication measurements. From the outset, the client needs to login to the Android gadget, utilizing the client account confirmation component. After a successful login, the client needs to check the QR code. The cloud-based administration stage transmits the drug data from the QR code on the prescription bundle to the smart medication acknowledgment gadget over a Wi-Fi network (Figure 7.11).

The smart machine acknowledgment gadget prompts a voice message to remind the clients to take the drug. At that point, the patient places the prescription in the acknowledgment area and presses the acknowledgment button to scan the medication. After the effective acknowledgment, the drug status (regarding whether the prescription is right, the prescription is inaccurate, more prescription should be taken, less medication should be taken, or other related medicine data) is reported to patient. The smart prescription acknowledgment framework transmits the results back to the cloud-based

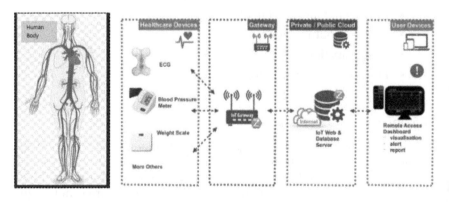

FIGURE 7.10
Smart healthcare.

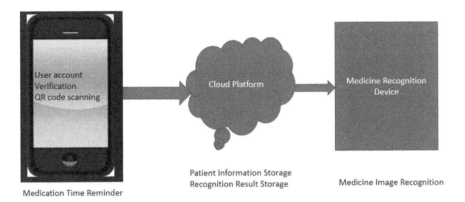

User account
Verification
QR code scanning

Cloud Platform

Medicine Recognition
Device

Patient Information Storage
Recognition Result Storage

Medicine Image Recognition

Medication Time Reminder

FIGURE 7.11
Drug recognition system.

administration stage over the Wi-Fi network. Therefore, relatives or the patient can check the patient's medical records (such as name, measurements and real prescription time) through the site.

7.4.6 IoT-Based Smart Environment Monitoring System

Environment is the surroundings or conditions of a geographical area. The environment greatly affects the life of living creature whether it be human, animals, birds or aquatic life. So, a good environment is the key to a good life. Environment monitoring is necessary in order to make decision about consumption of renewable or non-renewable resources. Environment quality and protection takes care of many issues like air pollution control, water pollution control, harmful radiation emission control, and so on. The IoT technology embedded with wireless sensors, cloud platforms and wireless communication technology play a vital role in environment monitoring and management. For environment monitoring, various types of sensor models are used. Each of the sensors perform a specific function based on the type of environment. The various models are:

1. *AirBot model*: The AirBot model monitors the airborne pollutants that can cause respiratory problem. This model is pocket-sized and is easily portable. The model was developed by Carnegie Mellon University.

2. *WaterBot model*: This model is utilized to measure water quality. One portion of the model is plunged into a water source and after that, it will transfer contamination information to the web through a ZigBee-introduced module. A sensor automaton model apparatus is likewise utilized that can detect numerous things in conditions, including gases, temperatures, humidity by cell phones.

3. *Lapka model*: This is a set of ecological sensors that can be attached to cell phones, particularly the iPhone, and can recognize electromagnetic radiation, nitrates in crude nourishments, temperature, dampness and even gives information to advise about natural nourishment.

4. *Air quality egg*: This model can be utilized in at-home conditions as a sensor pack. It gives readings of NO_2 and CO focused on any place it is put.

5. *Electronic nose model*: This is a multi-sensor gadget that recognizes limited quantities of perilous airborne synthetic concoctions like pesticides, burning emanations, gas leaks and substance-fighting specialists.

   • Air monitoring technique: Air pollution is a genuine worry in this day and age. It causes a worldwide temperature alteration and ozone layer consumption. Numerous sorts of gases, fluids or solids are scattered with standard air and make the air dirty which is a cause of weakening the strength of living creatures. Air contamination likewise causes acidic downpour. The acidic downpour contains destructive gases like nitric acid and sulfuric acid which is conveyed by downpour, haze and from the wind. The fundamental wellspring of air contamination is the consumption of petroleum derivatives like oil, fuel and coal; the ammonia gas produced from creature house, the misapplication of manures, herbicides and pesticides from rural activities; the outflow of unsafe gases from production lines or ventures; and mining activities.

IoT plays a key role in monitoring air pollution quality and reporting. Many types of IoT technologies are used for monitoring which include:

1. *Plume air technology*: This technology uses a personal wearable sensor which is used to report the weather forecast. The Plume air pollution monitor is attached with a phone or wearable monitoring device.

2. *Airy technology*: This is IoT-based hardware technology and uses LoRa technology for communication. It measures various pollution aspects like forest fire detection, traffic management modules, water quality monitoring and dust sensors.

3. *TreeWiFi*: This technology has been implemented in Amsterdam. It measures air pollution and makes the levels of air pollution visible through LED light. When the server detects an improvement in air quality, it allows the TreeWiFi system to share its internet connection with everyone on the street. Users that connect to the network get tips and tricks to improve air quality.

4. *ISPEX technology*: This technology is used to measure aerosols in the climate.

5. *Tzoa technology*: This is a wearable air quality and UV sensor innovation. It gets the client to learn about their ecological data through LED with a color-coded framework. It utilizes UV sensors to track sun exposure in this way dealing with a sound light level. Tzoa recognizes two kinds of particulate PM10 and PM2.5. PM10 alludes to particles of 10 micrometers or smaller, which will, in general, be allergens like dust. PM2.5 alludes to particles 2.5 micrometers or smaller, which incorporates the molecules found in vehicle exhaust, fine residue and smoke particles that are of moderate size and dangerous enough to cause genuine medical problems.

6. *Vehicular sensor network technology*: This innovation utilizes encompassing sensors to gauge the nature of the air. The VSN innovation conveys a progression of sensor hubs by open transportation and measures the nature of air on a specific geographic territory. Air contamination data is accessible to the general society through pages, web applications and portable applications. The sensor estimates the fixation O_3, CO and NO_2. The sensor hub utilizes the GPS module to get time and area data and Bluetooth connection to speak with a PC inside the vehicle. The information is stored away and dissected at the cloud stage. The framework gives profoundly exact air quality observation dependent on the IoT idea.

- IoT-enabled water monitoring: Water monitoring is essential, as water is an important element of human life. Water monitoring includes temperature monitoring, pH level monitoring, conductivity and salinity measurement. The remote monitoring of water is based on a wireless sensor network. Sensors in remote monitoring systems are integrated together and data records are stored and analyzed in a cloud-based storage unit. IoT ensures real-time management of water monitoring, providing the ability to optimize the use of clean water and managing water treatment plants to ensure the lowest cost and most effective service to customers (Figure 7.12).

- Weather monitoring IoT technology: Many technologies are used in weather monitoring such as satellites, radar, wireless technologies, sensors and hand-held systems.

  - Satellite system: Satellite systems integrated with IoT technology provide an ability to predict more accurate weather forecasts by observing from both geostationary and polar orbiting satellites. Upgrading the capacity of watch climate components on a worldwide scale and giving information about an area and transmission from remote perception frameworks,

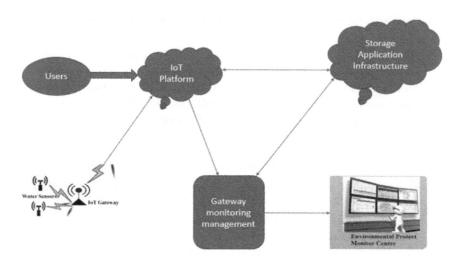

FIGURE 7.12
IoT-enabled water monitoring system.

the polar circling satellite pictures are particularly helpful for the recognizable proof of numerous highlights, such low clouds included as low stratus, haze and residue storms. These images are also useful in locating the presence of fires in a particular forest area. Satellite systems can be connected to internet networks to share the collected data to other monitoring database systems and with the benefits of the IoT, satellites will offer a great contribution to the environmental weather monitoring.

– Radar technology: Radar innovation has been a noteworthy segment in the location and warning of neighborhood extreme storms, including tornadoes and flash flooding. Radar items are created from both reflectivity and Doppler move data. Reflectivity information is valuable in creating items identified with tempest force, storm development and precipitation. Doppler shift information can help infer items identified with wind speed and shear just as with choppiness in the air, along these lines permitting the estimation of tornado development and other tempest structures. Numerous radar items are filled in as valuable illustrations to enhance the issuance of serious climate and flash flood alerts when dispersed in well-known media, for example, TV or on the internet.

– Hand-held technologies: Hand-held systems and sensors are used to collect parameters of weather like temperature, humidity, rain, snowfall, etc. Sensors are built in to smart mobiles which enable them to analyze the current weather conditions in a dedicated area. It is also supported by GPS to give an information of weather conditions about any region based on GPS coordinates. Small-scale Electro-Mechanical Systems (MEMS) innovation and micromachining strategies have been a well-known way to deal with the scaling down of sensors. The usefulness and unwavering quality of these smaller scale sensors has been expanded extensively by incorporating them with developing Integrated Circuit (IC) innovation or different sensors. This innovation is utilized to accumulate information, for example, temperature, dampness, gaseous tension, wind current heading and speed over a wide zone (Figure 7.13).

– Wireless technology: Wireless technology-based weather monitoring systems are an efficient way to monitor many weather conditions with a wide flexibility. Today, many wireless devices can be used to monitor the weather parameters offering a suitable data transfer through wireless like Wi-Fi, Bluetooth, WiMax, etc. A remote climate-checking framework empowers screening the climate parameter in an industry or any place can be structured utilizing ZigBee innovation. ZigBee is the most encouraging mechanical standard for low information rate and has a long battery life. Likewise, the ZigBee systems are reliable and self-recuperating

- Radiation monitoring: Radiation monitoring is essential to measuring the radiation emitted from different sources like nuclear

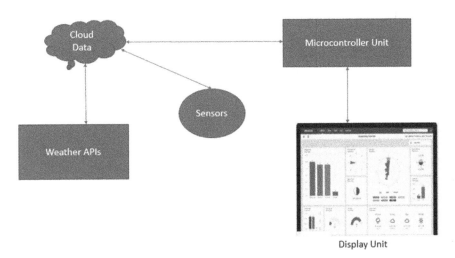

Display Unit

FIGURE 7.13
Weather monitoring system.

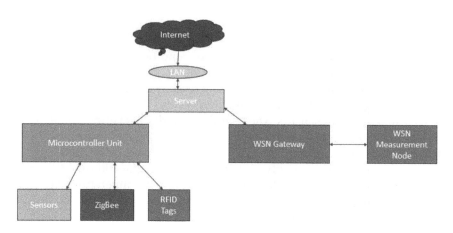

FIGURE 7.14
Radiation monitoring.

power, industrial areas and military sites. Different radiation systems are employed to measure radiation levels. The device must be capable of transferring data to highly sensitive systems. The system consists of many devices such as sensors, LoRa, ZigBee, RFID tags, mobile phones or WSN nodes (Figure 7.14).

7.4.7 IoT-Enabled Smart Energy Management System

Energy management refers to planning, controlling and reducing energy consumption. Today's organizations need to optimize their energy needs and drive their business without compromising their eco-sustainable growth. Increasing energy prices, emission of harmful gases and operation costs are some of the compelling factors to reduce, monitor, control and plan our energy consumption. The smart grid is one of the intelligent technologies for energy management. A smart grid enables new technologies to be integrated with the analytics to monitor and plan energy consumption. A smart grid helps users to know about their energy requirement, as well helping them to adjust their consumption according to requirement. Renewable resources like wind and solar power are sustainable and a growing source of electric power. The smart grid provides data automation needed and enable solar panel and wind to store energy on the grid and optimize its use.

A smart grid performs its operation in four divisions: energy generation, transmission, distribution and consumption. It consists of three types of networks:

- Home Area Network (HAN)
- Neighborhood Area Network (NAN)
- Wide Area Network (WAN)

HAN is the first layer and deals with consumers' power requirements. It comprises smart devices, home appliances, electrical vehicles and sustainable energy sources (such as solar panels). HAN is deployed inside private units, in modern plants and in business structures and connects electrical machines with smart meters. It is responsible for generation and transmission of power.

NAN is also known as Field Area Network (FAN). It is the second layer of a smart grid and comprises smart meters belonging to multiple HANs. NAN frames a correspondence between distribution substations and field electrical gadgets for power distribution frameworks. It gathers the data and services from numerous HANs and transmits it to the information authorities which interfaces NANs to a WAN.

WAN is the third layer of a smart grid and works as a base for correspondence between network gateway or aggregation points. It facilitates communication among power transmission systems, bulk generation systems, renewable energy sources and control centers. Also, a video camera has been mounted in smart grid management for video surveillance of appliances' safety, fire alarms and to monitor operations (Figure 7.15).

Thus, a smart grid improves reliability, security, safety and sustainability of the power system.

7.4.8 IoT-Enabled Intelligent Agriculture System

Food is the basic requirement of any human being and so is agriculture. Nowadays, smart cities are adopting IoT-based technologies for smart agriculture. In order to perform agricultural activities in smarter way, Web Map Services (WMS) and Sensor Observation Services (SOS) are integrated with IoT to check proper water management for irrigation. Agriculture IoT systems accurately monitor various parameters like greenhouse temperatures, soil moisture, weather and also integrates cloud-based recording systems. With IoT-based agricultural production, costs can be reduced to a remarkably low value which in turn increase profitability and sustainability. It is predicted by experts that by 2050, farmers will increase their production by 70%.

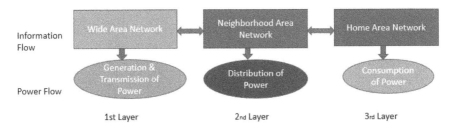

FIGURE 7.15
Smart grid management system.

7.5 Conclusion

The IoT-enabled smart city improves the quality of life and standard of living of its people. It is also helpful in economic development of a country. IoT-enabled management systems not only save money, but also save energy and time. Although we are moving toward the smarter city, the excessive use of any device is harmful. A smart city is surrounded by RF devices. The RF radiation emitted from the device is not suitable for the health of human beings. Also, the data in smart cities are stored on central cloud platforms, which are not secure and reliable. So, health and self-security should also be taken into account when developing a smart city.

Bibliography

www.intel.vn

Leonidas Anthopoulos. 2018. *Understanding Smart Cities: A Tool for Smart Government or an Industrial Trick?* Springer: Cham, Switzerland.

Wan-Jung Chang, Liang-Bi Chen, Chia-Hao Hsu, Cheng-Pei Lin, Tzu-Chin Yang. "A deep learning-based intelligent medicine recognition system for chronic patients", *IEEE Access*, 2019.

export.arxiv.org

www.iotcentral.io

www.treehugger.com

Bashir Muhammad Rizwan, Gill Asif Qumer. "IoT enabled smart buildings: A systematic review", *2017 Intelligent Systems Conference (IntelliSys)*, 2017. IEEE. London, UK.

M. Rogulski, A. Badyda. "Current trends in network based air quality monitoring systems", *IOP Conference Series: Earth and Environmental Science*, 2019.

A. Caperna et al. 2017. "Smart cities, local community and socioeconomic development: The case of Bologna." In *Smart Economy in Smart Cities* (Vinod Kumar, T. M., ed), Springer, Singapore.

B. Shwetha, B. Bhaskar. "Smart pesticide sprayer for Arecanut", *2017 International Conference on Recent Advances in Electronics and Communication Technology (ICRAECT)*, 2017. Publisher: IEEE. Conference Location: Bangalore, India.

Alex Grizhnevich. IoT for smart cities: Use cases and implementation strategies. *ScienceSoft*. Available at: https://www.scnsoft.com/blog/iot-for-smart-city-use-cases-approaches-outcomes

8

WoT-Enabled Retail Management

Deepti Mishra

CONTENTS

8.1 Introduction

Currently, the Web of Things (WoT) plays a crucial role in retail management. WoT has its roots in various domains where the retail industry is a key zone. It is reforming retail management rapidly by applying technologies for efficient functions. WoT-enabled retail establishments enable customers to access data easily when doing online shopping and allows them to efficiently manage their shopping cart. To offer better services to customers and retailers, we should focus on new technologies related to the retail industry.

Retailers also benefit, as they have the ability to manage the collected data to help customers to shop. It provides the retailer with the ability to understand, analyze and proceed with using web-enabled retail data for analytics.

WoT-enabled retail industry has the capability to manage enormous amount of data very efficiently and easily. Huge amounts of data that is gathered may belong to customers, retailers, products, reviews and stock, which further

147

require accurate data analysis. WoT provides proficient technologies to access, store, share and investigate a massive amount of data which is quickly generating on a daily basis. WoT offers data management with its investigations, predictive analysis and report generation with secure communication.

WoT contributes significantly in the supply chain and logistics in retail management to optimize time and cost.

The Web of Things is the combination of technologies which are connected to the internet in a web architecture with heterogenous software architecture [1]. It offers communication between various devices with the web. It focuses on software architectures, HTTPS, URLs services and communication protocols [2].

WoT provides a good infrastructure with the newest technologies enabled to offer proficiency to the retail industry both for consumers and retailers. WoT benefits allow them to optimize time and improve productivity.

The technology in retail management is evolving very speedily but still requires attention to issues like security and data management. It needs the focus on data flow, access control, secure communication, logistics and strong supply chain optimization.

The objective of the chapter is to present involvement of Web of Things in retail management. It also focuses on security issues which are currently faced by the WoT-enabled retail industry. The chapter presents the concept of the Web of Things with infrastructure, its applications and challenges in retail management.

8.1.1 Retail Management

Retail management is the chain of procedures through which customer benefits from various services directly, whether personal or commercial [3]. Figure 8.1 shows a conceptual process of retail management. It comprises of many factors such as consumer satisfaction, supply chain management, merchandising and promotions.

Retailing services can be store based or non-store based. Store based includes ownership of franchises, chain stores, supermarkets or small stores; non-store based are direct sales or e-shopping.

In many instances, the customer is unable to reach physically the store, and WoT-enabled retail management acts as crucial player on those occasions. While availing of WoT, retailers can know the time and location of customer for providing services by reacting proactively. WoT helps retailers to approach consumers rapidly by providing quality services easily, as they are integrated with RFID (radio-frequency identification) and the supply chain.

8.1.2 Advantages of WoT-Enabled Retail Management

The Web of Things offers numerous benefits to the retail industry. Both consumers and retailers gain from the involvement of the Web of Things in retail management in terms of profits and discounts on products [4]. Retailers

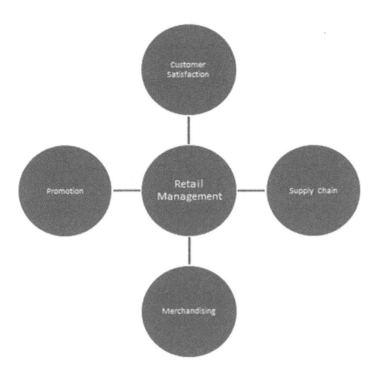

FIGURE 8.1
Process of retail management.

can track the stocks accurately to manage inventory. Consumers have lot of options and ease of shopping, as well as retailers and consumers both having better experiences with financial transactions [5].

1. No bargaining with customers, wholesalers and suppliers.

   Due to fixed prices for products, unnecessary wastage of time is prevented as is seen in the usual cases of retail shopping by customers' undue bargaining.

2. Demand attentive inventory.

   Current and future stock of products must be ascertained by the sale of products in abundance.

3. Raise in associated consumers.

   Every consumer is likely to increase in number of consumers in either arithmetic, geometric or harmonic progression depending upon the level of satisfaction.

4. Consumer and market analysis.

   Demand and supply are two key factors important in deciding the market trends. They are, however, also markedly influenced by the purchasing pattern of the customers, thereby helping to predict future market pattern analysis.

5. Simplifying store operations.

   The products which are not being sold in adequate quantities for a sufficient threshold period need to be replaced by those in high demand.

6. Smart store.

   A digital catalog of the store products helps in easy tracing of products, listing of shortages and early demand raises as per needs.

7. Increased productivity.

   Hassle-free working and regular updates with enough customer satisfaction are bound to give a higher productivity rate.

8. Home shopping for customers.

   It improves customer satisfaction. Customer is happy to receive products directly at home.

9. Rapid and efficient logistics.

   Timely delivery is nonetheless one of the most key factors of customer satisfaction.

10. Appropriate data analysis.

    Customer reviews supplement the market trends and thereby help in pattern recognition both for today and tomorrow.

11. Improved supply chain management

    Digital inventory maintenance with day-to-day data analysis helps in timely planning and arrival of stock shortages for prompt delivery.

8.2 Latest Trends in WoT-Enabled Retail Management

The market of the retail industry is mounting very rapidly through the involvement of the Web of Things. One such example is the fashion retail industry. The fashion retail industry had a boom after the involvement of WoT [6]. Nowadays, manufacturers are designing products in such a way so that user can monitor their health regularly. There are various applications of WoT in fashion retail management such as clothes with sensing devices, headsets measuring brainwaves, wearable devices to monitor heartbeats, devices to monitor sleep disorders, devices that count walking steps, infant monitoring and many more. These wearable devices are equipped with sensors, calculators and digital displays. Sensors are used to gather real-time data which are the physical parameters such as glucose values or blood pressure reading sensing from users. Then that data is calculated, and report is displayed on monitor attached with the device. Smart shirts are designed with respect to athletes for monitoring heart rate, stress levels and anaerobic threshold. Automated supply chains, Beacons, personalized discounts, reviews on products, smart stores, digital signage and data analysis tools are some examples of WoT-enabled retail management [7]. AWM smart shelves were implemented in California; Flonomics, a tool with a counting system with data analysis capability was applied in Denver, Colorado; and Engage3, an intelligent platform based on machine learning was implemented in Davis, California [8]. Smart warehouses, such as Flexe, were applied in Seattle. Digital Lumens, software based on machine learning is also in use [8].

WoT enhanced digitization both in stock management or in financial transactions. WoT helps the retail sector by providing better data analytics with the help of machine learning.

8.3 Infrastructure of WoT-Enabled Retail Management

WoT integrates customer, process, employee, manufacturer, object and services at same platform by digitization [9]. In terms of retail, the Web of Things includes Wi-Fi tracking systems, Beacons, RFID tracking system, digital signage, supply chain management and many more [10,11]. While designing infrastructure, retailers must ensure key points of applications such as automated checkouts, smart shelves, personalized discounts, offers and digital process optimization in implementation.

8.3.1 Beacons

Beacons play an important role in infrastructure designing of Web of Things-enabled retail management, as shown in Figure 8.2. Beacons are small battery-operated devices that transmit Bluetooth signals to nearby smartphones. Beacons reach out to customers that have Bluetooth-enabled phones. Beacons connect with the smartphone to send notifications, where the retail app has been downloaded, by displaying the information of sales on products on screen. Communication is one-way only, that is, messages can be sent by retailers through Beacons to customers on their smartphones, but the customers cannot reply.

Examples of Beacons are discussed showing their benefits for retail management.

- 10 June 2013: At the WWDC 2013 conference, Apple introduces iBeacon as part of iOS 7.

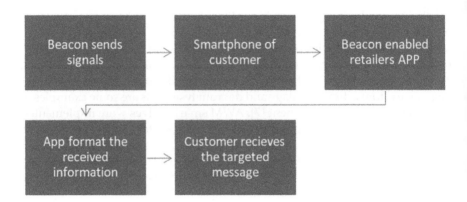

FIGURE 8.2
Functioning of Beacon in retail management.

- 1 September 2013: Titan installed Beacons on phone booths in Manhattan for maintenance purposes.
- 6 December 2013: Apple installed Beacon in 254 shops in the United States.
- 31 July 2014: Many US retailers test Beacons in their shops.
- 12 August 2014: Some UK stores installed Beacon for connecting with customers.
- 29 September 2014: Beacons were used by North America's retailers such as Macy's, Target, Urban Outfitters and CVS.
- 21 November 2014: Aruba networks an indoor navigation using Beacons.
- Early 2015: Some apps like Facebook started using Beacon in their functions.
- 14 July 2015: Eddystone was launched by Google which integrates its functioning with Beacon to provide better results in retail industry.

8.3.2 Eddystone

Eddystone transmits messages to nearby smartphones. It offers better outcomes, as it is a Bluetooth Low Energy Beacon, providing location and proximity-related functionalities. It can be managed remotely by integrating with Google services and can be sensed by Android and iOS. Eddystone's frame format includes Eddystone UID, Eddystone EID, Eddystone URL and Eddystone TLM.

8.3.3 Supply Chain Management Applying WoT

Nowadays, retail industries are restructuring supply chain management by applying the concept of the Web of Things. The functioning of supply chains improves when conjoined with the Web of Things as transfer of data takes place very efficiently [12]. The Web of Things drives the supply chain in an efficient way to trace and verify the products to expand quality in management. Functioning of WoT in the supply chain can be stated as it tracing products at any time by attaching devices to deliver timely messages. WoT helps the supply chain to locate products whether upward or downward and to monitor products in stock with scheduled management. It collects and makes data visible at crucial points, while concurrently managing more issues [13]. Real-time monitoring can play a crucial role in increasing the performance efficiency of supply chain management [14]. Inventory can be managed by prediction and future trends from the data saved from previous purchasing trends' seasonal and annual variations.

8.3.4 Web of Things in the Retail Industry

As the technology has advanced, it provides more benefits to retail industry. The infrastructure for a retail industry that is conjoined with the Web of Things includes hardware devices, communication, data management tools and applications for use.

Figure 8.3 shows the conceptual architecture of WoT-enabled retail management. The infrastructure is the integration of various physical devices

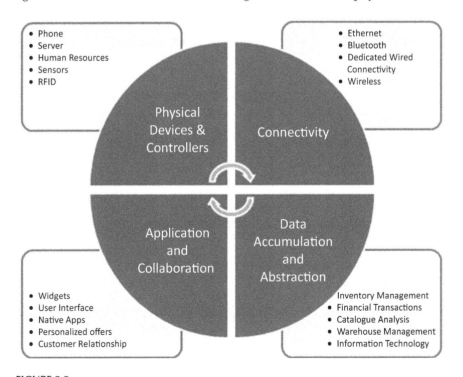

FIGURE 8.3
Conceptual infrastructure of WoT in retail management.

enabled with connectivity providing data analysis on various apps and desktops [10]. Hardware devices may be computers, cell phones, sensors, servers or human resources which further interact with each other at any end of the communication. Bluetooth or wireless connections are the basis for connectivity for WoT in retail management.

Currently, there are numerous tools for managing the enormous amount of data that is gathered in the retail industry for smooth processing, which are further connected and used by end users.

8.4 Applications of WoT-Enabled Retail Management

WoT offers many benefits and applications when used in conjunction with retail management system. A discussion of these applications follows.

1. Location Tracking

    Use of RFID makes tracking more reliable and accurate. It makes a digital record of arrivals and exits every time, thereby keeping the record of the details. The presence of tags makes item tracing quick, thereby enabling eliciting all the details including cost, discounts, expiry, etc. Barcode universalization helps in fetching the details of every product including batch, arrival, manufacturer, warranty and cost.

2. Inventory Management

    Products that are out of stock needs timely warnings and display to avoid discomfort to customers and embarrassment to service providers. Almost all products in one or other way require proper management techniques to avoid stocking up of excess or undue shortage. Live stock tracking is must on a priority basis as their quantity and quality may change without any warning in unfavorable circumstances. Patterns generated annually about purchasing trends and demand and supply, aided and abetted by general and local knowledge, helps in providing predictive accuracies regarding possible shortages forthcoming and thereby timely rectification measures.

3. Digital Payment

    Payment transaction reports help in timely dispatch of products as well as pattern generation for payment types preferred by customers along with data generation for daily transactions amount, as well as customer tracking in an inbuilt system. Nowadays, multiple payment gateways or e-payment solutions are in vogue which not only are easy to use for tech-savvy customers, but also cost effective for these customers by providing discounts and cashbacks, and are therefore very much preferred by them.

4. Interactive Digital Signage Screen

Visual searches on digital platforms encourages customer satisfaction about the type of product as well as comparison with other products instantaneously.

Default or local search engines are often voice-enabled ensuring exact and prompt search of the desired products.

5. Customers

Analysis of reviews of products provided by previous customers or by manufacturing companies help customers in decision-making. Price lists and capping help customers with limited budgets to easily find the suitable match as per their requirements. Product pricing remains variable as per demand and discounts and such tailored pricing helps distinct groups of customers to have their likes easily selected. Sites nowadays provide relevant search results in almost no time ensuring faster selection and quick purchases. E-commerce sites enabled with visual search and voice search have an edge over others lacking it, for reasons mentioned above.

6. Surveillance

After payment, customers can now have online tracking of products via messages from sites as well as from freight and logistic companies about the exact date, time and place of product delivery. Websites have specific domain and logins as well as specific redressal mechanism for tracing products, and reasons for delayed delivery or non-delivery and need for the same.

7. Real-Time Supply Chain

Synchronization of demand and supply needs extreme attention to avoid easily preventable shortages or excess stock pile-up and thereby losses and embarrassment on both sides. Analysis of annual records helps in sensing and predicting product demand based on customer purchasing trends. Proper and dedicated monitoring will ensure smooth functioning. Identical surveillance of logistics will ensure minimum grievance from customers by ensuring prompt delivery.

8.5 Security Issues in WoT-Enabled Retail Management

Online devices can be the great peril. The very famous Stuxnet attack almost sabotaged the Iran nuclear program by causing as many as a thousand uranium enrichment centrifuges to malfunction and finally fail [15].

International IoT facilities facing hacking incidents and IoT security issues have led to privacy violation, security beaches, business losses, infrastructure collapse and even health and medical emergencies [16]. The information provided by sensors should be in secure manner.

1. Breach of Privacy or Data

    There are always very heavy chances of data theft via network. The majority of them occur through third-party vendors. E-commerce firms are more prone to suffer from malware or ransomware attacks than those already secured; after every such incident productivity and faith of customers takes a huge dip. It is therefore mandatory for the enterprises to have prophylactic measures to prevent such mishaps.

2. Breach of Financial System

    Financial transaction security is of pivotal importance as these transactions are largely expedited from a distant location and the details are exchanged in the air [17].

    • Fraud and chargebacks: "Chargeback" refers to reversal of a transaction for consumer protection purposes from a fraudulent activity committed by merchants as well as by individuals. It is also a demand by a credit card provider for a retailer to make up the loss on a fraud or disputed transaction. Chargeback fraud, also known as friendly fraud, is when consumers fraudulently use it to get a refund. Consumers falsely dispute a transaction with the bank rather the merchant for a refund claiming that the product ordered was not delivered.

    • Cross-border transactions refer to in- and outbound transfers of property, stock or financial and commercial obligations between related entities resident in, or executing them in, different tax jurisdictions. As it occurs between two entities from different countries or geographical domains, it is also called an international transaction, ignoring territories or boundaries.

    • Card data security: In order to optimize the security of credit, debit and cash card transactions and protect card-holders against misuse of their data, a set of widely accepted policies and procedures are maintained, evolved and promoted by the Payment Card Industry Data Security Standard (PCIDSS) which is a global organization mandated by card brands and controlled by the PCI Security Council. Multi-currency credit card payments ensure payment in local currencies without involving foreign payment processors. You are paid in your own currency for the sale abroad in other currency. Multi-currency payment gateways allow merchants to offer their international consumers to pay in their own currency. With businesses reaching beyond country borders, such payments are the need of the hour.

- Technical integration is applied as a software suite combining a budget and accounting system that controls spending and payment processing, auditing and reporting, also called an Integrated Financial Management System (IFMS). It permits domestic investors to buy foreign assets and vice versa.

3. Over-Reliance on Technology

Nowadays, dependence on technology has increased for even the smallest task. Since the majority of task completions require works at multiple levels and interfaces, any small deficit or non-compatibility may lead to non-compliance or non-execution of the desired outcome. Many times we attempt to carry out a task which otherwise could be done manually/offline, the failure of which creates more mess. Failure also produces anxiety in users. There is a probability of failure during compatibility issues, such as monitoring with sensors' heterogenous software.

4. Complexity

Designing, developing and managing heterogeneous technology is a complicated task, as some techniques are long-standing and some are extremely new. Integrating new technologies with the existing ones is a barrier to executing the entire system efficiently. It is challenging to detect which device is functioning properly and which is not. It is difficult to figure out to detect topology without any manual related to WoT, as it is a combination of too many devices. Managing databases, applications, hardware and different software at a single platform is a tedious task.

5. Insecure Devices

Currently, devices which are included in the Web of Things for retail management are a major attack concern for corporate organizations. They require high demands of security methodologies as they can easily be attacked by unauthorized users or malware. Sometimes, lack of encryption techniques makes devices easy to hack. Smartphones are often victims of intentional data theft. There are numerous entry points in an organization's network and users' network, making devices insecure.

6. Tempering Communication

Connecting various communicating devices in real time for smooth and secure data transfer is a critical issue regarding connectivity [18]. During communication, gateways are most susceptible for interception. In the world of WoT, each adapter is connected to a new communication device, which can be further open for attack by hackers. Use of various protocols such as BLE, Thread and ZigBee in a single system increases the complexity of communication. Standardizing the protocols is a notified requirement in WoT-enabled communication. Identifying constraints on protocols during communication like congestion, missing packets, buffer overflow, wireless attack, slow processing and issues related to delivery of packets is key challenge.

7. Software Attacks

Web-enabled retail management can be easily attacked by any software malware, such as viruses or worms. Software attacks also include SQL injections for theft and to gain access to data. For rescue from such attack, management should validate the code and use prepared SQL statements. A denial of service (DoS) attack is a kind of attack to prevent authorized users from using services. For example, attackers may flood the network with large volumes of data. Data can be stolen by attackers as they may break cryptography techniques which are applied for data encryption.

8. Lack of Updates

Web-enabled management systems require automatic upgrading of devices regularly, which is again a big challenge for users and manufacturers. Lack of software updates has led to serious attacks by hackers [19]. Not applying regular software updates to a system leaves critical holes for attackers to steal private data. Additionally, software upgrading fixes security issues, enhances features, provides more protection from threats, and offers more compatibility with other devices. It should be applied on all devices including computers or cell phones. Therefore, it is necessary to upgrade the complete system in a timely manner to protect it from hackers.

9. Raw Storage of Data

Web-enabled retail management is mandatorily required for knowing the flow of data in the system, which is again a big security issue. It is very difficult to know how the data is created, the location of generation of data and the number of devices creating the data [20]. It is a laborious task to get the knowledge to capture data in real time for analysis and prediction. Usually, there is crude submission of data. That crude data is most vulnerable for attacks and theft. Modification in even small amount of data may impact the entire transferred information, which may be a further crucial loss.

8.6 Challenges Raised in WoT-Enabled Retail Management

The Web of Things is facing a lot of challenges in terms of infrastructure, security, competitors and data management [21].

1. Infrastructure

    Designing and updating the infrastructure for the WoT-enabled retail management is the key challenge. Many retailers and organizations lack the techniques required to design the infrastructure which is needed by the Web of Things. The Web of Things necessitates regular investment in digitizing by retailers such as RFID, tablets, networking devices and applications and data management tools [22]. Nowadays, new techniques are arriving fast, so it is a necessity to update existing ones. If retailers fail to update resources, they may lag behind their competitors.

2. Data Management

    To overcome this challenge requires applying techniques and software tools to analyze gathered data for improving the performance of the organization [16]. It is necessary to implement better techniques to collect, manage and analyze data for prediction and investigation for getting knowledge of future trends.

3. Security

    A long-term challenge for the Web of Things in retail management is handling security issues to protect the complete system. As numerous devices of both consumers and suppliers are linked to same system, security threats include theft of private data, breaking of encryption techniques, manipulation of information and hacking devices [23]. To cater to the challenge means applying intelligent techniques, not only for detecting security issues, but also predicting security threats in advance.

4. Threats to Competitors

    The Web of Things impacts the retail industry in various ways, such as reducing supplier bargaining power, which in turn increases threats to competitors [24]. Those retailers who are lacking in new technologies are facing the biggest challenge to remain in the market. Consumers' nature of purchasing and intensity of purchasing affect the supplier market.

5. Timely Service

    The goal of initiating the Web of Things in retail management is to improve timely and managed service. Service may be receiving products, resolving customer issues, backing up data, providing product reviews, accurate data searching so that customers can make timely decisions, etc.

8.7 Conclusion and Future Scope

Applying the Web of Things in retail can offer precious awareness to retailers in terms of profits and customers. It creates new opportunities and innovations in industries. Web-enabled retail management has designed new processes for customers for purchasing goods at their ease. This area has been enhanced a lot in terms of technology by including the functionalities of machine learning, artificial intelligence, data analysis and Web of Things. Still, it is facing a lot of challenges, both regarding customers and retailers, as it has not yet completely created by technologies. Security during a financial breach is the main concern while doing transactions. There should be robust systems to manage data privacy. It requires more robust techniques of encryption and security. Currently, there are various companies that have adopted WoT in their infrastructure to function smoothly and be on top of their competitors. Following this, it can be stated that the Web of Things has improved customer experiences as well as profits of retailers while improving retail management. With changing times and evolving newer technologies it is almost mandatory for entrepreneurs to have an edge over the others. So, to remain on customers' horizons and in vogue, WoT is truly indispensable for both sellers and buyers.

References

1. Mathew, Sujith Samuel, Yacine Atif, and May El-Barachi. 2016. From the Internet of Things to the Web of Things — Enabling by Sensing as-a Service. In: *12th International Conference on Innovations in Information Technology (IIT)*. Al-Ain, United Arab Emirates: IEEE. doi:10.1109/INNOVATIONS.2016.7880055.
2. Zeng, Deze, Song Guo, and Zixue Cheng. 2011. The Web of Things: A Survey (Invited Paper). *Journal of Communications* 424–438. doi:10.4304/jcm.6.6. 424-438.
3. Chen, Daqing, Sai Liang Sain, and Kun Guo. 2012. Data Mining for the Online Retail Industry: A Case Study of RFM Model-Based Customer Segmentation Using Data Mining. *Journal of Database Marketing and Customer Strategy Management*, 19, 197–208, https://doi.org/10.1057/dbm.2012.17.
4. Schwab, Klaus. 2017. *The Fourth Industrial Revolution*. Kindle edition, Penguin.
5. Khan, Rafiullah, Khan Sarmad Ullah, Rifaqat Zaheer, and Shahid Khan. 2012. Future Internet: The Internet of Things Architecture, Possible Applications and Key Challenges. In: *Proceedings of the 2012 10th International Conference on Frontiers of Information Technology*. 257–260. doi:10.1109/FIT.2012.53.
6. Nayyar, Shaminn. 2019. The Impact of Internet of Things on the Fashion Retail Sector Bringing Experience to Retail. doi:10.13140/RG.2.2.32135.04008.

7. Barthel, Ralph, Andrew Hudson-Smith, and Martin de Jode. n.d. Future Retail Environments and the Internet of Things (IoT). doi:10.13140/RG.2.2.36396.56963.
8. Thomas, Mike. 2019. No Checkout Lines, Personalized Shelving and the IoT Retail Revolution. April 11. https://builtin.com/internet-things/iot-in-retail-tech-applications.
9. Dziak, Damian, Bartosz Jachimczyk, and Wlodek J. Kulesza. 2017. IoT-Based Information System for Healthcare Application: Design Methodology Approach. *Applied Sciences*, 7(6), 1–26, 596 doi:10.3390/app7060596.
10. Sethi, Pallavi, and Smruti R. Sarangi. 2017. Internet of Things: Architectures, Protocols, and Applications. *Journal of Electrical and Computer Engineering (HINDAWI)* 2017: 9324035. doi:10.1155/2017/9324035.
11. Guinard, Dominique, and Vlad Trifa. 2016. *Building the Web of Things: With Examples in Node.js and Raspberry Pi*. Shelter Island, NY: Manning Publications.
12. Phase, Avani, and Nalini Mhetre. 2018. Using IoT in Supply Chain Management. *International Journal of Engineering and Techniques* 4(2), 973–979.
13. Kumar, Amit, and Omidreza Shoghli. 2018. A Review of IoT Applications in Supply Chain Optimization of Construction Materials. In: *34th International Symposium on Automation and Robotics in Construction* 471–478. doi:10.22260/ISARC2018/0067.
14. Kothari, Sneha S., Simran V. Jain, and Abhishek Venkteshwar. 2018. The Impact of IoT in Supply Chain Management. *International Research Journal of Engineering and Technology (IRJET)* 5(8), 257–259.
15. Sicari, S., A. Rizzardi, L. A. Grieco, and A. Coen Porisini. 2015. Security, Privacy and Trust in Internet of Things. *Computer Networks: The International Journal of Computer and Telecommunications Networking* 76: 146–164. doi:10.1016/j.comnet.2014.11.008.
16. Roman, Rodrigo, Pablo Najera, and Javier Lopez. 2011. Securing the Internet of Things. *Computer (IEEE)* 44(9). doi:10.1109/MC.2011.291.
17. Mathew, Sujith Samuel, Y. Atif, Quan Z. Sheng, and Zakaria Maamar. 2013. The Web of Things—Challenges and Enabling Technologies. In: *Internet of Things and Inter-cooperative Computational Technologies for Collective Intelligence* (Bessis, N., Xhafa, F., Varvarigou, D., Hill, R., Li, M., eds). Springer. doi:10.1007/978-3-642-34952-2_1.
18. Sun, Wencheng, Zhiping Cai, Yangyang Li, Fang Liu, Shengqun Fang, and Guoyan Wang. 2018. Security and Privacy in the Medical Internet of Things: A Review. *Security and Communication Networks*. doi:10.1155/2018/5978636.
19. Greengard, Samuel. n.d. *The Internet of Things*. Boston, MA: MIT Press Essential Knowledge Series.
20. Mora, Higinio, David Gil, Rafael Muñoz Terol, Jorge Azorín, and Julian Szymanski. 2017. An IoT-Based Computational Framework for Healthcare Monitoring in Mobile Environments. *Sensors*. 1–25. doi:10.3390/s17102302.
21. Baker, Stephanie, Wei Xiang, and Ian M. Atkinson. 2017. Internet of Things for Smart Healthcare: Technologies, Challenges, and Opportunities. *IEEE Access (IEEE)*. 5, 26521–26544. doi:10.1109/ACCESS.2017.2775180.
22. Singh, Kalpana. 2014. Retail Sector in India: Present Scenario, Emerging Opportunities and Challenges. *IOSR Journal of Business and Management (IOSR-JBM)* 16(4): 72–81.

23. Velasco, Carlos A., Yehya Mohamad, and Philip Ackermann. 2016. Architecture of a Web of Things E-Health Framework for the Support of Users with Chronic Diseases. In: *Proceedings of the 7th International Conference on Software Development and Technologies for Enhancing Accessibility and Fighting Info-Exclusion.* ACM. 47–53. doi:10.1145/3019943.3019951.

24. Boora, Krishan Kumar, and Kiran. 2016. Assessment of Five Competitive Forces of the Electronic Retail Stores in India: Expansion and Growth of Modern Retailing. *IOSR Journal of Business and Management (IOSR-JBM)*, 18(11), 30–34. doi:10.9790/487X-1811063034.

9

WoT-Enabled Banking Sector Modernization

Aarti Jain

CONTENTS

9.1 Introduction

The banking sector, which is the core of financial services, is a necessity for the socio-economic growth of any country. The banking sector has existed in every country for ages, doing its work to boost the economy. But in this era of digitization and internet, the banking sector is constantly challenged by the GAFAs (Google, Apple, Facebook and Amazon) and by the companies offering/developing the various financial software applications [1].

As indicated by Capgemini's World Fintech Report of 2017, 50.2% of clients state that they preferred digitization and automation and have thus started opting for the banks with more IT-enabled facilities [2]. Thus, for any bank, in order to keep itself alive in the financial market, there is a requirement to endorse the upcoming technologies so that it can be more client-driven and create more advanced facilities to its customers [3].

The fundamental task of any banking sector in an economy is to play its part in the growth of the financial service sector by transferring the money from the bodies with surplus capital to those bodies who need these funds. Most of the decisions about lending money by banking sector are based on the assumption that borrower will pay back the loan amount with interest. For the individuals who deposit their money in the banks, banks facilitate them with easy deposit policies, security in respect of their money, interest on the deposited money and easy access to the deposited money [4,5].

WoT or the Web of Things is no longer a matter of research only; it is now implementable. WoT is changing the manner in which we perform certain tasks and its availability is making businesses smart, cheaper and more and more customer friendly. WoT mainly works on the principle of using data analytics to automate the tasks which were initially tedious for humans to do efficiently [6]. For core banking requirements, such as Know Your Customers (KYC) norms fulfilment, credit lending, risk management, non-performing asset calculations, recovery, collateral management, finance trade and various types of insurance, WoT has smart solutions to deal with them. Combined with other developing advances, for example, digitization of the whole work, WoT can make new peer-to-peer (P2P) plans of action that can possibly improve banking in a couple of zones. WoT is the interconnection of remarkably recognizable inserted registering gadgets inside the current internet framework. WoT is relied upon to offer availability of gadgets, frameworks and administration that goes past machine-to-machine (M2M) interchanges and covers an assortment of conventions, areas and applications [7]. In the banking sector, the interconnection of these WoT-enabled services with digitization will introduce automation in many banking-related services.

WoT has the potential to completely modify the way the banking sectors are working. As the banking sector mainly deals with huge money transactions, these money transactions generate huge amounts of data, which is collected and analyzed for various banking sector-related operations. WoT has a strong grip on this data, which in turn can benefit both the banking sector and the customer [8]. WoT is the biggest technological evolution which is claiming the biggest ever digital revolution in future. WoT technology in the banking sector will help customers save time, investing smarter and planning a more appropriate lifestyle, and to the banking sector to increase its revenue by offering better customer services. WoT in the banking sector is still in its infancy, but there is huge scope for its growth and further innovation in the WoT-enabled banking sector.

This chapter is organized as follows: Benefits of WoT in the banking sector; examples of WoT in banking and financial services; challenges and design issues with WoT-enabled banking services; and growth rate of WoT-enabled banking services.

9.2 Benefits of WoT in the Banking Sector

The banking sector is moving toward the digitized world to avail of, as well as offer, numerous better approaches to serve its customers and to gather significant information of customers through various sensor-based gadgets. By adopting digitization through the means of WoT, the banking sector has effectively implemented M2M communications to users' screens and remote frameworks. Here, the remote framework implies CCTV, digital signage and ATMs. WoT innovation has also enabled the banking and funding industries to help their customers innovatively and help them accomplish superior business results [9]. Thus, WoT has turned out to be a significant advancement leading to banking innovation. Directly from upgrading client experience, streamlining activity to set up computerized banking machines to diminish line time, acquiring adaptability administration and increasing the efficiency of representatives, WoT has turned into a piece of "things to come in banking." The next section presents the benefits of WoT in the banking sector.

9.2.1 WoT-Enabled Product Planning and Customized Marketing

With the WoT innovation, banks can launch better services and help centers for its customers. With the help of information gathered by WoT, the banking sector can plan what service or offer can benefit both the customer as well as the banking sector. Furthermore, it will help the banking sector to choose the perfect time for launching new services [10]. Customized marketing is the most ideal approach to hold the client in an aggressive market. WoT innovation has made it feasible for the bank to monitor all customer exercises and offer items and administration according to their necessity.

WoT can assist the banking sector in understanding the present economic situation of the customer and accordingly the banks can plan to offer various services and investment schemes to its selected customers. This will guarantee keeping up a solid association with their client and a better client experience.

Today, the banking sector is investing more and more into reliable services, promoting its business, offering better services to its clients and automating various approvals and verifications for fast delivery of banking services. This vision of today's banking sector can be formalized by using cloud services. In cloud-based services, banks are required to store their clients' information on the cloud and can utilize this information to offer fast and better services. One such example of cloud-based banking is by using customer relationship management (CRM) programming, which is a client relationship with the board and CMS, which is a content administration framework. Using these two frameworks and some decision rules, the customer archives are stored on to the cloud in such a way such that they are secure and easily available to banking system. All this data and its implications can be used to improve business tasks and secure customers' recent financial transactions [11].

9.2.2 WoT-Enabled Proactive and Prediction-Based Banking

WoT-enabled banking services have made it feasible for the banking sector administration to identify any kind of administration or financial issue. Also WoT-enabled banking services bring notice of any discrepancy and forward it to the concerned in good time, such that the bank can easily resolve the issue. With WoT innovation, the bank can follow and check past data and client conduct before offering any banking-related service. This way, any unusual activity or action in the client's record can be identified and analyzed. This client-related information in WoT-enabled banking can be obtained from social media, e-commerce websites and through portable applications and advanced technology. Today, every bank has portable and sensor-based applications for banking like ATMs and pay-in slip generators that also give information about the client. This information can also help the banking sector analyze client conduct and prerequisites and helps in automating and reliable banking transactions [12].

9.2.3 WoT-Enabled Credit Credibility

The profit earned by banks is based on the credit-related services, such as loans like home loans, motor-vehicle loans, gold loans and personal loans, which it offers to its customers. For banks, these services need a guarantee that the borrower will not be a defaulter. Often banks take special measures to check the client's capability to repay the loan, but even then, many of the loans are not paid back and result in losses to the banks. The creditor data/information assessment via social media, ecommerce websites and a wide range of sensor-based gadgets have a lot of effect on the reliable credit business of banks and will have to continue to aid the bank throughout the loan period for timely repayment of loan. The sensor-based gadgets which are the gadgets perceiving the present business status of customer, medical coverage and living style, the bank can approve the credits/loans to its customers in a hassle-free manner. With the help of these gadgets, the bank can also decide on the backup plans, in case of any default by creditors, and improve strategic approaches. This backup plan basically focuses on the customer's actual wealth, spending behavior and possible income increments to assess the customer's hazard [13].

9.2.4 WoT-Enabled Bank Data Management

Banks have to process a large set of data and investigations on a regular basis and they need to screen every one of the money exchanges to maintain a strategic prevention from any unethical services. With the help of WoT, data can be made accessible to the bank and can be utilized by various frameworks and systems of banking. For example, a customer who is using its banking app on mobile has an alternative to click a snapshot of the required document and upload it onto banks server by using mobile services. These user-friendly services will increase profit and productivity and improve the customer's trust and relationship with the bank.

From the bank's perspective, for example, the bank advances home loans to its customers after proper verification and documentation. On these home loans, the banks can also extend maintenance loans after a few years if required by borrower. However, for disbursing the maintenance loan, physical verification of the required maintenance has to be done by the bank. For automating this service, banks which are utilizing WoT innovation can give their borrowers a choice to install sensors in their new house. These sensors provide reports to the bank in advance about any required home improvement or maintenance. These installed sensors can also help in expediting the approval process and will reduce desk work and fieldwork [14].

9.2.5 WoT-Enabled Credit Card Facilities

A reliable network connectivity can give important client data and information to banking-related administrations. After analyzing customer data like shopping choices, love of gadgets and travel choices, the banking administration can then inform its selected customers about new offers in a shopping mall, travel locations, credit quantification, things to buy from nearby places, places to visit nearby, and so on, by using the client's current location. Generally, low-range communication, such as Bluetooth or Wi-Fi technology, is preferred for sending notifications to these selected customers. This is a very innovative way to advertise for people who are really interested in a particular product [15].

Because of the advent of WoT-enabled services in banking sector, it offers customized, solid and fast services to their clients. This has been possible due to the data which has been collected and analyzed when it is required.

WoT has the capacity to accelerate the setting of the interface between the banking sector and its clients and enables them to address the client's issues in a fast and reliable manner.

9.2.6 WoT-Enabled Banking-Related Information Dissemination to Customers

WoT can help banks in educating its customers about various customer-related banking practices. For instance, when people need a loan for a vehicle or house, utilizing WoT, they can check the up-to-date information regarding the procedure to be followed for obtaining the loan and loan repayment schedule. If their loan is approved, the customer can check their credit balance, premium paid, principal paid, prepayment if done, current interest rates from any point without physically visiting the bank. This example is only one facility which can be linked with WoT-enabled banking services [16].

9.2.7 WoT-Enabled Personalized Customer Services

Taking into account that inclinations and choices of the client are dynamic in nature, banking organizations focusing on development in different financial solutions and banking innovation cannot disregard new advanced digital developments and hence are required to participate in the improvement of the new digitized world. With the advancement of technology, clients today keep contact with their bank by using mobile phones, a tablets or personal computer. This accessibility of banking services to the client through digital technology, through which banks are collecting information about clients, their conduct and inclinations aids the banks to observe and propose further improvements required in WoT applications. The association of banking services with WoT has enabled the gathering of large amounts of information on the client's conduct as well as their future financial interests. This information about clients enables banks to serve its clients in a more professional way and offer them more specific and customized services thereby developing a fruitful bond for both. Hence, it enhances a new level of understanding and connection between bank and its customers, which further leads to business enhancement and further innovation. For example, imagine a client enters a bank branch from the main entrance gate. The bank's WoT-enabled system can utilize various biometric sensors and programming to validate the client's identity with the help of face recognition. Similarly, a client may usually withdraw money from the ATMs of your bank. In this case, WoT-enabled services can guide the ATM to get the cash ready for the client and also guide the customer in its preferred way of withdrawal like denominations and amount. Furthermore, banks can offer various face-recognition-based security services to its customers [17].

9.2.8 WoT-Enabled Customer Smart Interaction

For now, capacities of the WoT in the banking sector are not being fully explored. WoT has initialized its first step into the financial sector. However, it is evident that WoT will play a significant role in banking methodologies. Specialists have foreseen the added specialties of WoT-enabled banking services where clients will have access to the option to link their bank-related services to its personalized gadgets. For instance, smart devices like watches can raise an alert to its owners, when they spend more than an amount predefined by the customer. Today, apps developed for the banking sector send an SMS/alert to the customer, whenever the customer spends money on a service or product. However, with the advent of WoT, these banking applications will also suggest a better alternative spending plan.

Banks throughout the world are attempting to utilize WoT abilities in their industry to include more clients. Beginning from portable banking applications that are utilized today by almost every bank in the world, since applications help in monitoring the interest of clients, to the scope of sensors that make it workable for banking sector to assemble data from their branches and sensors in wearables that can monitor how clients use banking items [18].

Example: With the different types of sensors and programming techniques, cashless payments can be by customers without contacting client's mobile phone or any banking card. For instance, clients may simply enter into the sales section of a store, and sensors implemented there will automatically identify the type and quantity of items in the basket. Based on the type of items, the total cost of the items is calculated and payment is made through a customer mobile wallet. This mobile wallet may use face-recognition sensors to verify the client's identity.

9.3 Real Applications of WoT in the Banking Sector

There are numerous WoT-enabled applications, which are made available by several banks to its customers. A few of these applications are as follows.

9.3.1 Banking on Wearable Devices

Due to low-cost wearable devices, many banks have started offering their services by developing applications for the same. Smart watches like Apple Watch have already been in use under the banking system of Bank of America. Banks have also started launching their own wearable gadgets for smart banking like "pay with BPAY" by Barclays [19]. Other banks that have started offering their services on wearables are Hellenic Bank and Australia's WestPac.

9.3.2 Chatbots

Chatbots have some appropriate uses in retail banking. Chatbots are a simple and generally moderate method for mechanizing client assistance enquiries. Some new businesses have pushed on this open door as well; for example, the Cleo application interfaces with your bank and Facebook's chatbot to assist customers with their queries. Also, there are other chatbot-based models from the Royal Bank of Scotland, such as RBS Assist chatbot for banking. This chatbot answers FAQs to RBS utilizing Kasisto's Kai AI stage and enables clients to directly exchange their views and take care of payments. Swedbank, in the interim, propelled Nuance's "NINA" on its site, a versatile application to help answer client requests all the more rapidly by sourcing the pertinent data [20].

9.3.3 Alfa-Bank-Sense

This is a WoT-enabled bank application which predicts regular client finance-related conduct and offers the item or administration the client may require around then. This WoT application is more customized than some other versatile banking application and conveys information like Facebook Messenger does [21].

9.3.4 ATMs on the Move

With the advent of digitization and WoT, banks have come out with unique solutions, which are ATMs on the move. Recently, Idea Bank launched ATMs on the move services by running a fleet of cars, each occupied with a security system and an ATM. These cars visit the client, instead of the other way around. The bank's information indicates that with these ATMs on the move, the average deposit rate is three times higher than the traditional style of depositing. In the interim, in Canada, the credit association Blueshore is investigating the possibility of displaying banking-related app notifications and information on vehicle windshields, for travelers to audit their portfolios while traveling [22].

9.3.5 Blockchain and WoT-Based Banking

Blockchain technology, which has the capability of maintaining a secured and protected record of confirmed exchanges is being abundantly talked about, in the banking sector and beyond. A few banks are, as of now, trialing the innovation. These include the Commonwealth Bank of Australia, Wells Fargo and the trading firm "Brighann Cotton," which have guaranteed to complete a main worldwide exchange between two banks utilizing block-chain, smart communication and WoT [23].

9.3.6 Voice-Assisted Banking

Voice-assisted banking is one of biggest advancement in banking systems. Recently, Capital One, a US-based bank, made it feasible for clients to take care of their payments by means of Amazon's Alexa. Similarly, a UK challenger bank Starling, for instance, has tried different things with Google Home Assistant. It has coordinated its APIs with the speaker, which empowers the clients to ask questions about instalments through voice directions.

9.4 The Challenges Associated with WoT-Enabled Banking Services

In this section, we will discuss the other side of WoT-enabled banking services, such as the challenges associated with the WoT-enabled banking services. The biggest potential challenges with WoT-enabled banking services are accuracy and security of data collected by WoT devices. The banking sector which is using WoT should ensure the security and privacy of individuals' information for their clients. A few of the potential challenges are discussed in detail as below.

9.4.1 Data Privacy and Security

With the developing web-based banking, data security and data privacy are the noteworthy challenge for any banking sector. Since WoT is a network made up of different types of gadgets, which are further using different programming platforms, WoT-enabled banking services are prone to data hacking and data manipulation. Further, wireless communication and long-distance communication made WoT-enabled banking services more vulnerable for data hacking. Moreover, this issue of data privacy and data-hacking is riskier for banking services as it deals directly with financial transactions. Thus, for implementing WoT-enabled banking services, data privacy should be of high concern. Thus, WoT innovation ought to guarantee that the entire banking system is protected and secure.

For instance, when a client associates their groceries intake with WoT-enabled bank services, the WoT-enabled pantry system automatically senses whenever any grocery item falls low. In response to that, the WoT-enabled pantry orders that grocery item from the local market. In respect to this purchase, the WoT application for a money transfer to the local grocery store owner runs in the background. As here we are dealing with a client's bank account, the client must have some safety measures to evade any deceitful exchanges.

Within a few years, WoT will control all money transactions and end clients will have the option to do nearly everything whenever it might suit them, from any place and whenever through the web. It offers new chances to take care of issues and increase the value of monetary administration to its clients.

9.4.2 Standards and Protocols

Various gadgets require different types of support systems. Unfortunately, there are no common standards, or even architecture, for developing hardware as well software for WoT-enabled services. The main reason behind this issue is that there are number of manufacturers and developers working in this WoT area and they have their own specified protocols and standards for WoT. Moreover, if all WoT device manufacturers in the world agreed by consensus to utilize one explicit protocol/standard, technical issues would still exist. The only solution is that, for one kind of service, only one manufacturer should be there, which is practically and financially unrealistic. Hence the lack of common standards will hamper the growth of WoT.

9.4.3 System Complexity

WoT-enabled services are the Web of Things that are communicating with each other. The number of such devices is very high and the system they are dealing with is very complex, too. Moreover, WoT-enabled banking services have direct connections with the economy and financial stature of any country and thus failure of this system will result in huge losses.

WoT is basically a chain of devices, applications and software, where each connection assumes a significant part of completing the task assigned. For successful implementation of WoT-enabled banking services, there is requirement of high-quality hardware manufacturers and brilliant software developing companies. Hence, though WoT in the banking sector is advantageous, it is still a challenging task for proper implementation.

9.4.4 Higher Unemployment Rate

WoT has mechanized the working procedures that required human brains. It implies that for a given specific job, lower numbers of people are required than before. Consequently, some employees will lose their jobs, especially younger employees and the one who are not specialists. Banks will eliminate more positions as WoT advances in the financial sector. This decrease in employment rate will harm the social and economic prosperity of any nation, as well as of the entire world.

9.5 Conclusion

From the discussion, it has been proven that WoT has the potential to modify the banking sector. In order to enjoy the benefits of WoT, it is now the sole responsibility of the banking organizations to come forward, adopt this emerging technology and accept it with full confidence. WoT has the capability with which the banking sector can be expanded globally and can benefit different countries with their mutual financial growth. WoT is an emerging and powerful area, but before adopting any technology for implementing WoT, it is mandatory to invest a lot in security systems to protect financial data.

The data collected after adopting the WoT technologies will add value to the financial sector and will help in implementing globally accepted standards. WoT will help the banking sector to look ahead and predict various financial aspects beforehand. It is now up to the banking sector whether they accept the growing WoT technologies to take a step forward in the financial world or want to pause its evolution.

References

1. Akinci, S., Aksoy, Ş., & Atilgan, E. (2004). Adoption of internet banking among sophisticated consumer segments in an advanced developing country. *International Journal of Bank Marketing*, 22(3), 212–232.
2. Didenko, A. (2017). Regulating FinTech: Lessons from Africa. *San Diego International Law Journal*, 19, 311.
3. Riemer, K., Hafermalz, E., Roosen, A., Boussand, N., El Aoufi, H., Mo, D., & Kosheliev, A. (2017). *The Fintech Advantage: Harnessing Digital Technology, Keeping the Customer in Focus.* University of Sydney, Business School and CapGemini.
4. Sathye, M. (2003). Efficiency of banks in a developing economy: The case of India. *European Journal of Operational Research*, 148(3), 662–671.
5. Rajan, R., & Dhal, S. C. (2003). Non-performing loans and terms of credit of public sector banks in India: An empirical assessment. *Reserve Bank of India Occasional Papers*, 24(3), 81–121.
6. Guinard, D., & Trifa, V. (2009, April). Towards the Web of Things: Web mashups for embedded devices. In: Proceedings of WWW (International World Wide Web Conferences) Workshop on Mashups, Enterprise Mashups and Lightweight Composition on the Web (MEM 2009), Madrid, Spain (Vol. 15).
7. Zeng, D., Guo, S., & Cheng, Z. (2011). The Web of Things: A survey. *JCM*, 6(6), 424–438.
8. Barnaghi, P., Sheth, A., & Henson, C. (2013). From data to actionable knowledge: Big data challenges in the Web of Things. *IEEE Intelligent Systems*, 6, 6–11.

9. Tornjanski, V., Marinković, S., Săvoiu, G., & Čudanov, M. (2015). A need for research focus shift: Banking industry in the age of digital disruption. *Econophysics, Sociophysics and Other Multidisciplinary Sciences Journal (ESMSJ)*, 5(3), 11–15.

10. Nicoletti, B. (2014). Mobile banking. In: *Mobile Banking* (pp. 19–79). London: Palgrave Macmillan.

11. Cravens, D. W., & Piercy, N. (2006). *Strategic Marketing* (Vol. 6). New York: McGraw-Hill.

12. Bilal Zorić, A. (2016). Predicting customer churn in banking industry using neural networks. *Interdisciplinary Description of Complex Systems: INDECS*, 14(2), 116–124.

13. Soman, D., & Cheema, A. (2002). The effect of credit on spending decisions: The role of the credit limit and credibility. *Marketing Science*, 21(1), 32–53.

14. Codd, E. F. (1970). A relational model of data for large shared data banks. *Communications of the ACM*, 13(6), 377–387.

15. Newkirk, M., & Newkirk, L. (1984). U.S. Patent No. 4,439,636. Washington, DC: U.S. Patent and Trademark Office.

16. Jayachandran, S., Sharma, S., Kaufman, P., & Raman, P. (2005). The role of relational information processes and technology use in customer relationship management. *Journal of Marketing*, 69(4), 177–192.

17. Shamdasani, P. N., & Balakrishnan, A. A. (2000). Determinants of relationship quality and loyalty in personalized services. *Asia Pacific Journal of Management*, 17(3), 399–422.

18. Gibbert, M., Leibold, M., & Probst, G. (2002). Five styles of customer knowledge management, and how smart companies use them to create value. *European Management Journal*, 20(5), 459–469.

19. Mehrnezhad, M., Ali, M. A., Hao, F., & van Moorsel, A. (2016, December). NFC payment spy: A privacy attack on contactless payments. In: *International Conference on Research in Security Standardisation* (pp. 92–111), Springer, Cham.

20. Adams Becker, S., Cummins, M., Freeman, A., & Rose, K. (2017). *NMC Technology Outlook for Nordic Schools: A Horizon Project Regional Report*. Austin, TX: The New Media Consortium. Cover image courtesy of BigStock Photography, 978–1000.

21. Cuomo, S., Di Somma, V., & Piccialli, F. (2018). Pricing estimation of a barrier option in an IoT scenario. *Future Generation Computer Systems*. doi.org/10.1016/j.future.2018.01.027.

22. Acharya, A., Li, J., Rajagopalan, B., & Raychaudhuri, D. (1997). Mobility management in wireless ATM networks. *IEEE Communications Magazine*, 35(11), 100–109.

23. Memon, R.A., Li, J.P., Nazeer, M.I., Khan, A.N. and Ahmed, J., 2019. DualFog-IoT: Additional fog layer for solving blockchain integration problem in Internet of Things. *IEEE Access*, 7, pp. 169073–169093.

10

The Revolutionary Future Impact of the Web of Things in the Retail Industry: Toward Intelligent Retail

Amit Bhati and Ashish Gupta

CONTENTS

10.1 Introduction

10.1.1 Retail Management

Retail management is one of the latest parts of sales management. It is for the most part centered around upgrading the client's experience through various procedures to make all boosts imaginable for making a purchase. Retail managers center around building up an engaging environment using color, assortment, sizes, room temperature, shelf height and width, product position or group contributions among numerous different systems are utilized to set up the most ideal stage for more deals to happen. Accessibility is likewise a pivotal component tended to by retail management.

Stores should put in a lot of effort to ensure that everything the customer needs is appropriately shown and customer support agents ought to be able to rapidly provide whatever the customer is requesting. This training has been demonstrated to be entirely productive for organizations and it is presently generally actualized. Huge brands that sell retail products arrange certain spaces with retail foundations to show signs of improvements in visibility and enough space to show the product appealingly. Additionally, they utilize individuals to direct the manner in which products are set, to ensure they look as engaging as possible to fill in as components of influence for customers passing by.

10.1.2 Digital Technology

The most recent 30 years of innovative improvement has brought people and social orders into the digital age. The internet, and the ensuing reception of mobile technologies and gadgets, has influenced all domains of our everyday lives. The entrance to learning has multiplied through the presentation of new channels, similar to social network sites. On the one side, this ought to help people settle on better educated choices. However, on the other, it has created a condition of entropy, with a lot of market decisions and little to pick between the competitors.

In this condition of entropy, sentiments, rationale and qualities – the human angles – utilize the gigantic amount of digitally created information to settle on decisions and activities. In any case, digital innovations don't simply open unlimited sources of knowledge.

Perhaps, in the coming future, these digital technologies may even make decisions using artificial intelligence. When we look at the effect that computerized innovations have had on our lives, and our everyday basic decision-making process, shopping is a fine example of exactly how things have changed in recent years. In the pre-internet period, the buying of garments was fundamentally the same for practically all buyers. They could peruse design magazines, decide on a thing they loved and afterward share their perspectives with their family and companions. They would then go to the store, where they would try on various things with the assistance of a shop collaborator. Purchasers could then purchase the item or leave the store to go to another. Today, be that as it may, utilization of the internet and smartphones has opened up new information channels for buyers. Purchasers presently end up looking at an assorted, rich and prompt progression of data about the items of attire they need to purchase. This flow of data tails them any place they go. Utilizing their smartphone, they can see items, read reviews, share conclusions, start with one online shop then move onto the next and buy the item. What's more, they can do this any place they are, and at whatever point they like.

With digital advances, shoppers become progressively mindful of their potential outcomes. Thus, they need to customize their shopping background. They need to stand out from the group. Yet, they additionally need to use the power and pervasiveness of innovation for increasingly convenient and engaging methods for shopping.

10.1.3 Customer Experience in Retail

Desires concerning customer experience, administration, speed and proficiency play an unmistakably increasingly significant job for the present channel-rationalist and associated portable buyer. While logistics can assume a key job in gathering these requests and giving a better customer experience and administration (for example, by utilizing track and follow with full perceivability for the anxious shopper), there are numerous different regions where the Web of Things (WoT) can help improve customer experience as we will cover. From computerized signage and self-checkout to savvy mirrors, they all can assume a task.

According to a published report in 2016 [1] on IoT (Internet of Things) venture information, IDC Ltd. stated that in-store logical showcasing will be one of the quickest developing cross-industry WoT use cases somewhere between the year 2017 and 2021 with a 20.2% Compound Annual Growth Rate (CAGR). It clearly will assume a noteworthy job in retail too.

10.1.4 A New Age of Retail Management

To understand the revolution in the retail sector let us consider a scenario. A customer strolls into a store looking for newly launched cell phones. As he enters, his existing cell phone pings, and he open it to discover a guide, demonstrating where the new range of cell phones are placed in the store. He strolls over to them, give them a shot and put them in his basket. A robot moves up to him and engages him, inquiring as to whether he needs assistance discovering anything else. The robot takes him to a power bank showcase, as he has mentioned, and discovers his choice. When he has finished shopping, he leaves the store right away. His items are then scanned by sensors as he leaves and the final cost is then taken from his mobile payment application or his Wi-Fi debit card. Since he has shopped there recently, he receives an automatic discount. The rack where he got the power bank, in the meantime, observes the purchase and sends that data to a back-end stock framework, so the retail location's director knows to re-stock.

Sound excessively like a smart home? Possibly; however, this experience is nearer than one might suspect, because of the ascent of WoT, which makes a system between web-associated physical gadgets. In the following couple of years, physical gadgets fit to be associated with the web will keep on rising. The situation above is clarified from a client's point of view; however, it's imperative to observe exactly how profoundly WoT can influence retail location proprietors and representatives. As per Manyika et al. in 2016, McKinsey & Company [2] gauges that the potential monetary effect of IoT in retail situations will run from $410 billion to $1.2 trillion every year by 2025. WoT can diminish stock mistakes, upgrade the retailer's inventory, network the executives and reduce work costs. At last, WoT can support their conventional physical shop in contending with the present online-first shopping world, by exponentially improving the client experience and diminishing superfluous costs.

10.1.5 Motivation

The fundamental objective of this chapter is to present the connection between the Web of Things and retail businesses. The general objective is to break down the job of the Web of Things in the enhancement of business and to investigate how WoT innovation can positively affect e-retailing as well as negative reaction to WoT. This examination recognizes the opportunities and the difficulties of WoT in retail organizations.

10.2 Literature Review

10.2.1 Technology and the Retail Business

As revealed by Grewal et al. [3], electronic development can engage purchasers to get increasingly engaged and supportive shopping offers, secure speedier organization and at long last, make increasingly educated decisions. Then again, because of imaginatively made efficiencies, retailers accomplish reasonable purchases at discounted costs. The two groupings have profited at the same time, which finally improves the organization's advantage. All these innovative advances have as an objective increasing the client conduct, and also consumer loyalty. Receiving electronic trade methods and consolidating them with investigation, a retail location can quicken its presentation and reinforce the client commitment. On the off-chance that a retailer incorporates these advancements with modern rising forces, it can accomplish a high evaluation of intensity. One of these forces is the Internet of Things. In a "smart world," IoT has the main job permitting clients and retailers to coincide agreeably behind a cloak of robotization and furthermore, shrewdness. So as to be increasingly explicit, Grewal et al. [3] complies with the last-mentioned idea, showing the "smart homes" which are furnished with insightful machines that can see the amount of goods they have and solicit another order when it's beneath the typical level.

As indicated by Abazi [4], the last innovation presents us with another sort of information use through which organizations accumulate data and use them to inspect, measure what has increased and wrap up a specific issue which affects their business. There is an incredible association among IoT and clients. The more the SMEs are related with this innovation, the more possibilities they have to get data about clients their clients and increase their satisfaction. However, Abazi [4] states that the primary concern that leads associations to have an aggressive edge is the change for development. Not every one of the organizations have grasped this new technological pattern, giving a head start to different organizations that have involved it in its activity. Those organizations that have learned to be creative and versatile with the new computerized changes have increased incomes in comparison with organizations that have faded into still using out-of-date innovations. At long last, as per Abazi [4], IoT will impact distinctive business zones, while around 25% are starting at now using the Internet of Things advancements and have an incredible effect chiefly on the assembling divisions. Notwithstanding this, one of the basic guideline concentrations for an association is to extend its abilities concerning data for the Web of Things.

Singh and Singh [5] predicted that by 2020 in excess of 30 billion machines would be associated with the web and enabled by sensors. Organizations will take advantage of the interconnected gadgets by opening new ways to its activities and using all the significant information that they offer. Altogether, IoT gadgets suggest an extended capability and the upsides of this can be firmly connected to the business. The above authors note that through this innovation organizations can use the information that is created from the gadgets and have valuable insights of knowledge through the analytics which can prompt high purchaser fulfillment and streamlining of business procedures. Moreover, the brilliant devices can help with efficiency by making requesting without anyone else being involved, and along these lines save time and use these advantages in the best conceivable way. IoT can likewise improve the stock system and the inventory network by coordinating them and giving constant detectable quality. At long last, one of the fundamental advantages of IoT is the improvement of well-being and security which can be accomplished through recognition systems and smart sensing.

Iansiti and Lakhani [6], alluding to the GE case, show that organizations can completely change their plan of action because of the advancement of the

Internet of Things and the entire digitization process. The digitized transformation can influence the plan of action in two ways: On the formation of significant worth to the customers, and on the way, they benefit from that. The availability of the articles can change the organizations from multiple points of view. The adjustment of advanced sensors to the enterprise's products, their relationship to a cloud programming stage, the allotment of cash-flow to new programming applications and the improvement of their examination limits could bring about an expansion in the company's incomes because of their items' expanded proficiency. In addition, this change can lead to high aggressiveness which can be obvious on exchange forms, on the investigation and the social event of information just as on the correspondence among individuals, products and activities.

10.3 Challenges in Retail Management

10.3.1 Retailers Need to Contribute More for the Workforce

As indicated by the US Agency of Labor Statistics, starting in 2016, more than 15 million individuals are employed in the US retail industry. Likewise, the activity of a retail partner calls for something beyond accomplishing sales and being the administrator of the store. With the consistently evolving situation, retail associates nowadays should advance from being only "salesmen" to being "specialists and experts." In spite of innovation progressing at a rate of knots, retailers won't depend on the most recent circulation models, varied methodologies or mechanical advances to win the market. The ones who are triumphant in the market will accomplish it by conveying an important human communication that will offer confidence to shoppers in what to buy. This requires an expanded responsibility in employing, preparing and enabling the business partners to consistently put client cooperation first. In more straightforward words, a retailer planning to remain focused in the present market needs to bring its A-game with regard to employing staff and building up the staff's range of abilities. Gone are those occasions when preparing a worker on products and store strategies were sufficient. The need to prepare the staff to relate better to the customers is crucial. Connecting with the customers is basic, if any retail business is to grow.

10.3.2 A Siloed Showcasing Framework Makes It Costly and Inconvenient to Get Message Across

Engaging with clients crosswise over various channels is necessary for organizations in today's advertising world. From messages, to SMS, to web-based life, the multichannel interchanges are fundamental for engaging with clients, as this is the thing that drives the production of an excellent client experience. Nonetheless, with such huge numbers of isolated channels, it isn't so uncommon for client information to move toward becoming siloed. In the event that all the moving pieces of an advertising division are not successfully conveying and working together, clients can move toward becoming overpowered with clashing or rehashed messages. This flood of advertising interchanges can, without much of a stretch, have something contrary to the expected impact on clients and result in clients moving to competitors with a far clearer message.

The best way to handle this issue is to guarantee that one has the correct innovation and correspondence methods set up. All arms of an advertising group must be in agreement and have an unmistakable procedure for what they need to accomplish. A reasonable procedure will enable the retailers to set up every one of the channels in cooperation, instead of neutralizing each other. This won't just guarantee that the retailer arrives at the client with an unmistakable message, however; it will likewise save money and time.

10.3.3 Keeping Up Customer Loyalty

One of the fundamental factors in creating brand devotion is guaranteeing a good client experience. A 5% increase in client retention can expand a company's profitability by 70%. Retention is an immediate consequence of a faultless client experience, as both are characteristically connected. The most widely recognized missteps that retailers make are losing their current clients and imagining that they can be effectively replaced by the new ones. In the event that a retail business keeps this mentality, it will discover extraordinary trouble in continuing its business development. While offers and promotions contribute to helping clients feel like they are extraordinary, the most significant viewpoint to a complex background is personalization. Becoming acquainted with clients based on their interests and past buys can enable retailers to drive faithfulness. These bits of knowledge can be harvested from information, or even a straightforward discussion. In spite of the fact that most aspects of harvesting these pieces of knowledge will rely upon the size of the business, nobody ought to be too enormous for a quick talk with a standard client. A basic customized message and offers can be conveyed to the clients through their favored method for contact. Indeed, even a straightforward customized email can improve things greatly. Envisioning what the client needs ought to be the bleeding edge of a business, as it will assist the business with ushering the clients down the business conduit toward their next buy.

10.3.4 Clients Moving to Multichannel Purchasing Encounters

Innovation has been a noteworthy boon to retailers, and to clients as well; in any case, they additionally assume a noteworthy job in pushing the difficulties in the retail industry. Keeping pace with innovation will be perhaps the best challenge for retailers in the near future. With innovation quickening at a cosmic rate, retailers must keep pace if they are to remain important. Customers need an encounter to match that of Amazon and Alibaba – one that is fueled by Artificial Intelligence (AI), Machine Learning (ML) and huge amounts of information.

With more e-retail encounters accessible and delivery times decreased radically, it isn't that difficult to think why over 90% of Americans utilize internet shopping somehow or another. Online retail increased 300% somewhere in the range of 2000 and 2018, as indicated by the US International Trade Commission. Retail establishment deals dropped practically half during a similar period. In the initial three months of 2019, 5,994 stores closed their activities, contrasted with 5,864 during the entire year of 2018. Be that as it may, these are similar Americans who still spend a noteworthy segment of their allover shopping spending plan in the customary brick-and-mortar areas. In other words, despite the fact that everybody is shopping on the web, they are still making the greater part of their buys in person.

As customers are seamlessly moving along on the web and disconnected stages for shopping, they are ending up progressively open to retailers which can best encourage these exchanges. With the explosion in versatile retailing, in-store research and showrooming are more typical than at any time in recent memory. On the other side of the coin, online shipments can be conveyed to a nearby store, further closing the gap among disconnected and online retail.

The solution here is to focus on providing an unparalleled experience for clients over all of the channels. Clients are consistently in the lookout for retailers they can trust to convey unmatchable service on numerous occasions. Just having the correct client information can help the retailers in making an omnichannel experience for the clients, enabling them to associate any place and any way they wish.

10.4 WoT Applications in Retail

10.4.1 Automated Checkout

Retailers have likely perceived to what extent lines dissuade their customers from buying items. Also, as an administrator, it can feel unprofitable to pay various staff members to work during quieter shopping times. With WoT, a retailer can set up a framework to read the labels on everything when a customer leaves the store. A checkout framework would then count the items up and consequently deduct that cost from the customer's mobile payment application. Making a computerized checkout framework utilizing WoT would make retail industry customers more pleased and all the more ready to enter the store, particularly in the event that they are on a time crunch. According to Manyika et al. in 2016, McKinsey & Company [2] showed automated checkout can decrease clerk staff necessities by up to 75%, bringing about reserve funds of $150 billion to $380 billion per year in 2025.

10.4.2 Personalized Discounts

In the event that a retailer often has returning customers, they would like to reward them for their loyalty. With WoT, they can set up sensors around the store that send discounts to specific customers when they remain close to the product with their cell phones, if those customers pursue a loyalty program ahead of time. Also, they can utilize WoT to track the status of a product a customer has been looking at on the web, and send that customer a customized rebate when they are coming up. Envision in the event that customers examined a bathing product on the web, and afterward, coming up, received a discount on their preferred bathing product. As opposed to offering general discount on a wide assortment of products, the retailer can tailor each limit utilizing WoT to boost their transformation rates. Eventually, discovering approaches to consolidate WoT technology into their everyday business requires imagination and foreknowledge, yet the advantages of WoT in retail as sketched out above can assist a retailer's business to find creative arrangements with attracting important and loyal long-term customers.

10.4.3 Smart Shelves

A lot of retailer staff time and vitality is centered around monitoring the products to guarantee that they will never be out-of-stock, and watching that products will not be lost on different racks. A retailer can utilize smart shelves to automate both of those tasks, while at the same time preventing potential theft. Smart shelves are fitted with weight sensors and use RFID labels and tag scanners to filter the items on both presentation and stock racks. Smart shelves notify the retailer when items are running low or when they are inaccurately put on a rack, which makes their stock procedure savvy and increasingly exact. Moreover, each RFID tag is associated with a tag scanner, so smart shelves can identify in-store theft, saving money on security staff and cameras.

10.4.4 In-Store Layout Optimization

A retailer may be shocked to discover his retail space is not upgraded for their customers' behavior – possibly their least well-known products are in the front, or customers would lean toward more space around the loveseats in the back. By utilizing passageway examination software with infrared sensors, a retailer can utilize WoT innovation to improve their retail format.

Maybe the retailer discovers the majority of their customers invest most of their energy looking at TVs, yet those TVs have been placed in the back of the store, behind seldom-sought DVD players. This data arms the retailer with significant customer behavior learning, so they can put things customers care about most, such as TVs, in the front of the store.

10.4.5 Optimizing Supply Chain Management

While retail stores can already track products without the assistance of WoT technology, the data following it is really restricted. With RFID and GPS sensors, retailers can utilize WoT to receive progressively exact information, such as the temperature at which an item is being put away, or the amount of time it spends traveling. Retailers can utilize that information to improve the nature of transportation pushing ahead – and, even better, they can likewise act progressively if an item is being kept at temperatures that are excessively low or excessively high, avoiding a considerable misfortune.

According to a TATA Consultancy Survey and Business Insider, producers using IoT arrangements in 2014 saw a normal 28.5% expansion in incomes between the years 2013 and 2014. In the event that they have a long queue of providers, truck drivers and merchants dealing with their items, it is basic for them to precisely monitor how their item is taken care of and where it is situated in the store network. This data encourages them to guarantee their procedure is running as effectively as could be expected under the circumstances, and can help retailers to get their item into their customers' hands as quickly as possible.

10.5 Challenges in Retail Management for Implementing WoT Technology

10.5.1 Infrastructure

Most retailers do not have the infrastructure and system segments that immense volumes of IoT data require. With the end goal for traders being to digitize their retail locations, they would need to have a strong system, cloud solutions and end-user solutions, for example, standardized QR/barcode scanners, tablets and mobile point of sale (mPOS). Those things would require impressive investment.

The solution here is that there is no compelling reason to overinvest in infrastructure at the same time with regard to actualizing another innovation. You can begin with little infrastructure changes, for example, utilizing IoT to manage AC or the lighting, which will bring an increasingly prompt ROI. You can, bit by bit, get increasingly advanced with your IoT solutions.

10.5.2 Security and Privacy Issues

Numerous retailers are careful about the security and privacy issues related with IoT. These worries have been intensified by the presentation of GDPR (General Data Protection Regulation). Access to the client's information gives retailers different chances and yet opens the entryway to digital assault dangers and legitimate difficulties.

Retailers should work intimately with IoT programming engineers to ensure that the gadgets and sensors they use are structured in light of solid security instruments, including essentials like secure passwords, and further developed security framework like start-to-finish encryption, standard programming updates and an IT foundation that effectively filters for bugs and vulnerabilities.

10.5.3 Data Management

Completing IoT data analysis in a timely and relevant way speaks to a huge challenge for retail businesses because of an absence of pertinent capabilities and skills. There isn't sufficient specialized and scientific expertise close by to increase profitable experiences from the enormous measure of data gathered from IoT.

Retail organizations can contract area specialists or rely upon outsiders with the applicable IoT capabilities and preparation, who can assume control over information the business gathers. By addressing those difficulties, retailers get an opportunity to make their IoT investment productive, while increasing their focused edge in the market.

10.6 Conclusion

Each and every day retailers are discovering new techniques to attract customers to buy their products and make more profit. In this article, we have addressed the advantages of WoT in the retail sector, along with explaining how technology can play an important role in the retail sector by making retail management easier. Currently, numerous retail companies are adopting WoT technology to better serve their customers along with strongly competing with their market competitors. Case studies showed in this article have proved that WoT has not only improved the buying experience of customers but also gives good growth in retailer businesses in terms of time, money, and so on. Today, with the use of machine learning and artificial intelligence technology, WoT devices can demonstrate the products, and can also perform comparisons to similar products along with suggesting what kind of products a customer will need in the future based on their shopping habits.

References

1. IHS Technology. 2016. *IoT Platforms: Enabling the Internet of Things*. Paris, France: IHS Technology.
2. McKinsey & Company. 2015. *The Internet of Things: Mapping the Value beyond the Hype*. New York, NY: McKinsey & Company.
3. Grewal, D., Roggeveen, A. L., and Nordfalt, J. 2017. The future of retailing. *Journal of Retailing* 3(1): 1–6.
4. Abazi, B. 2016. An approach to the impact of transformation from the traditional use of ICT to the Internet of Things: How smart solutions can transform SMEs. *IFAC-PapersOnLine* 49(29): 148–151.
5. Singh, S., and Singh, N. 2015. Internet of Things (IoT): Security challenges, business opportunities & reference architecture for ecommerce. In: *2015 International Conference on Green Computing and Internet of Things (ICGCIoT)*. IEEE.
6. Iansiti, M., and Lakhani, K. R. 2014. Digital ubiquity: How connections, sensors and data are revolutionizing business. *Harvard Business Review*, November, p. 18, https://hbr.org/2014/11/digital-ubiquity-how-connections-sensors-and-data-are-revolutionizing-business (Accessed: 22 September 2019).

11

AndroSet: An Automated Tool to
Create Datasets for Android Malware
Detection and Functioning with WoT

Manju Khari, Renu Dalal, Udit Misra and Ashish Kumar

CONTENTS

Alarming growth rates of malicious applications pose a grave issue that has set back the robust mobile ecological community. A novel study analyzed that every 10 seconds, a new malware application is introduced in Android. To counter this serious malware campaign, we require a scalable malware detection approach that can efficiently identify malicious apps from a pool of both harmful and benign apps. Innumerable amounts of tools for malware identification have been introduced at system level as well as network layer. But here in this work we have proposed an automated tool that would extract features from the Application Package Kit (APK) files and form a dataset of it, which would then be used for static analysis of the data. All the research papers based on Android malware detection claim to have made a dataset of various features like permissions, intents, API calls, and so on, but none of them categorically state the procedure to form a dataset from the APK files. In our work, we focus mainly on the proposed automated tool that would create a dataset by reading the APK files and extracting features mainly from two files, for example, Manifest.XML and Classes.dex. After having this tool, researchers can create their own dataset and they won't have to rely only on few datasets available in the market for analysis. The only thing required for our tool is an APK file. The whole extraction process to forming a proper dataset in a Comma-Separated Value (CSV) file would be done by this proposed automated tool. The tool is created in the Python language.

11.1 Introduction to Android

Smartphones, e-books and most mobile platforms extended their over-taking wings so rapidly that they have become pervasive because of their immensely personal and dynamic qualities. The current trend shows that mobile shipments had surpassed personal computers in 2010 [1], which thereby stimulated an advance of sophisticated malware mobile applications. In 2014, McAfee acquired 6 million mobile malware samples and approximately 98% of them were based on Android devices primarily [2]. Due to Android's ubiquitous performance and the susceptibility to hazards of the mobile Operating System (OS), it must have effective approach to support the development of identification and analysis of malware applications.

To showcase market security issues, the two most predominantly used mobile platforms (Google Android and Apple iOS) use different approaches. On the one hand, Apple undergoes a method of manual inspection of all the applications submitted to the App Store before they are published. This intervention allows the Apple employee to read the description of the application and ascertain whether the resources and information used by the application are appropriate. On the other side, Google does not perform any inspection before adding the application to the Play Store. It mainly depends on a permission-based system for security purposes. Application developers must request the permissions from the individual users to access secure and private information and resources. However, at the time of installation, the whole permission list is displayed to the end user with the assumption that the user is able to figure out the listed permissions are appropriate.

There are three major reasons found for the rapid increase of malware applications in the Android market, which includes unsatisfactory documentation, substandard developers and malicious nature [3]. Moreover, a study also showed that the number of permissions in the application increased rapidly in the years between 2009 to 2011, and mainly in malicious categories [4]. Various studies have been undergone and in many, it has been found out that malicious applications tend to ask for more permissions than benign applications. From the million apps received by Androids [5], between 2010 to 2014, it was found that malicious applications needed approximately 12.99 permissions, while reliable applications required only 4.5 permissions.

TABLE 11.1

In 2017, Sales of Operating Systems in Smartphones Worldwide (Thousands of Items)

Operating System	2017 Items	2017 Market Share (%)	2016 Items	2016 Market Share (%)
Android	1,320,117.1	85.93	1,268,562.73	84.78
iOS	214,923.5	14.03	216,064.04	14.38
Other OS	1,494.01	0.10	11,332.22	0.82
Total	1,536,536.3	100.6	1,495,959.12	100.01

According to a report by Gartner, nearly 86% of smartphones have Android as the operating system, as shown in Table 11.1 [6]. Dependency on Android has risen rapidly and thus there is a need to protect users' interests by providing a secure application free from malware or any form of threats.

Recently, more efforts have been put toward analyzing behavioral data of Android applications and other online data processing systems [7–9]. The apps requested permissions that to some extent help to indicate its functionalities as well as its behavior during its runtime. So the machine learning approach and the data analysis approach have started pervasively to find out the malicious application. DREBIN [10] became one of the most prevalent tools which used machine learning algorithms with static data analysis to predict malware applications. It showed that the result can be further increased by incorporating more features if possible to aid detection. So our proposed tool is capable of automated extraction of all the permissions and features of each application and making a dataset. It does this by extracting the APK file, which is used for the distribution of Android applications as a standalone package. The APK is an archive that contains the source code for that specific Android application, so the features can only be extracted by deciphering this file. After extracting the features, we can then apply different Data Analytic techniques to the dataset generated by the proposed tool, to predict malicious applications [11–14].

In 2008, the Android freeware operating system was released and was known as Android. It is built on the topmost layer of a Linux kernel. Mostly, Android applications are based on the Java programming language and also on the recently released Kotlin language by JetBrains. The source code for these applications is run on Dalvik Virtual Machines. The Android architecture as shown in Figure 11.1 can be divided into six major layers: the kernel at the bottom layer, low-level tools, the hardware abstraction layer, C/C++ libraries, the Android Runtime and the framework of Java API and all applications on the top layer. For core system utilities like management of memory and process, security, network stack and driver model, Linux version 2.6 is used for Android. Between the hardware and software stack, the kernel layer works as an abstraction layer [15]. A standard interface provides the hardware abstraction layer (HAL) which reveals the capabilities of a hardware device

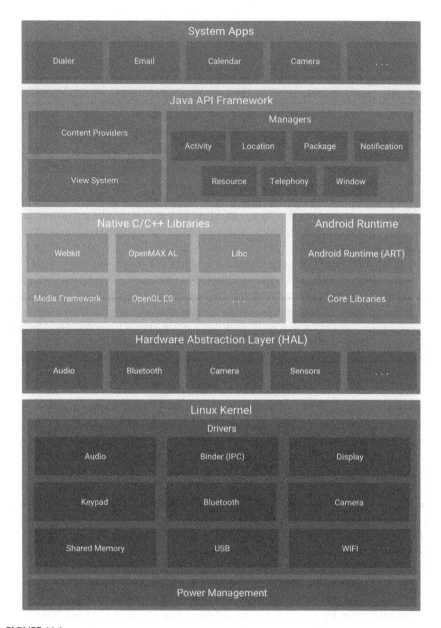

FIGURE 11.1
Android architecture.

to the upper-level Java API framework. Each application works in their own process and their own instance of Android Runtime (ART), when running in Android version 5.0 with API level 21 or above. DEX files are executed on virtual machines with minimum memory devices; these files use a byte code format for the Android which optimized the minimum memory footprint.

Services and components, like ART and HAL, are used in the core Android system; they are created from novel code which needs native libraries that are using in C and C++ language. Java framework APIs provide the Android platform to reveal the utility of a few native libraries to applications. Due to the Java language, the Android OS is easy to access for end users by using API. Android has the features of core application for SMS end users, email, internet browsing, calendars, contacts and many more. The platform included many apps which have no significance between the apps the end user chooses to install. These are the six layers which form the core of Android.

An Android application consists of four major components, that is, activities, broadcast receivers, services and content providers. An activity is a single screen of an application like a settings page or a browser window [16]. Activities are the actual visible components of Android. Broadcast receivers are used to send or receive messages between various components. The broadcast receivers are initiated through the *onReceive()* method and are invalidated on return from this method. Services are used to run background tasks in Android. A service permits an application to execute long-running tasks in the background and allow some of the application's functionalities to other applications in the system. Content providers are used to manage structured data stored in a local database powered by SQLite. These providers can be used to share data among multiple applications, provided that the involved applications have the correct permissions to access data. All these components are using intents which are used as a medium of communication between various components. Every Android application is distributed in a package known as APK. These APK consists of the source code and various resources required by the application. There are basically two types of analysis possible for malware detection in Android.

- Static analysis
- Dynamic analysis

Static Analysis: Static analysis is used to analyze the source code of the application without actually executing the code. There are two types of static analysis possible, namely [17], signature-based and anomaly-based detection. Signature-based is used to determine the SHA associated with the application and to determine if the application has the same SHA. Signature-based detection can be easily bypassed by changing only a small subset of code and is thus not very effective. Anomaly-based detection is used to detect anomalies in the source code of the application. The actual source code of Android applications is not readily available and thus we need to extract all the code from APK files. The APK files in Android consist of the following components.

1. META-INF: This contains various pieces of meta-data associated with the Android application like RSA, SHA-1 digests, list of resources, etc.

2. Assets folder: Ts used to hold files which can be retrieved by the Android application.

3. AndroidManifest.xml: This is used to keep the references to the major components used in the application like activities, services, broadcast receivers and content providers along with their intents. It also contains the permissions asked for by the application.

4. Classes.dex: These are the dex files which contain the actual source code of the application. It contains the various API calls performed by the application from the Android SDK.

5. Lib folder: This contains the compiled code which is specific to the hardware, that is, the processor implementation of the Android device.

6. Res folder: This contains the resources for the Android application which are not compiled.

The main files to consider in the APK for static analysis are AndroidManifest.xml and Classes.dex. The AndroidManifest.xml contains the permissions asked for by the applications which are the key component in detecting the malicious applications. It also contains various intents which can be used to determine if the specific component performs any malicious operations. Classes.dex contains the actual compiled code which is run on the Dalvik Virtual Machine. Thus, we have to first decompile it and then look for any API calls which point to the application being malicious.

Dynamic analysis: Dynamic analysis requires the code to be executed and then it observes the results obtained from it. It only shows one path per execution, but this can be improved by simulation. There can be many triggers which execute a service, or send an intent because of highly interactive nature of Android, and thus a new layer of complexity is added, because of the malware being "live" interacting with the environment. The two approaches for analysis can be combined so that the malware detection is enhanced even more. This approach is known as hybrid analysis. The approach is quite robust and can increase code coverage to find vulnerabilities.

11.2 Components of AndroidManifest

1. Permissions: Permissions are an important feature for the detection of malware in Android systems. For example, if the Android application wants to read data from the device's external storage, then it has to specify that it needs permission by mentioning it in

AndroidManifest.xml. The framework checks if the specific permission is there or not and only then will it grant the application access to the external storage. There are two types of permissions:

- Normal permissions
- Dangerous permissions
  - Normal permissions: These are the permissions which are granted by the system by default as these permissions do not cause harm to the user's privacy.
  - Dangerous permissions: These are the permissions for which the user is asked at runtime, as these permissions can cause harm to the user's privacy.

2. Intents: Intents are used to communicate between the components of Android and are thus important for the analysis of malicious applications. The private data of the user can be leaked using intents. The intents of an application need to be declared in the manifest and hence we can perform a static analysis over the intents.

3. Hardware components: Hardware components also require permission from the user. For example, if the application needs to use GPS, then this permission also has to be specified inside the AndroidManifest.xml by the developer and thus we can also make use of this as a feature in our dataset. For example, a calculator application would not need to use GPS or 3G.

11.3 Use of Android in WoT

It was once a concept considered to be in the realm of science fiction. Instead, the "Web of Things" (WoT) is already right here and increasing at a fast rate. WoT refers to the growing interconnectedness of various smart gadgets on the internet. All these gadgets have characteristic sensors and the net connection allows the devices to collect and send data. Extensive types of gadgets already exist in the market and lots of others are ready to be launched as developers move toward expanded connection of electronic gadget in houses and workplaces. Now, without difficulty, one may manipulate the refrigerator, treadmill, smart TV or toaster by using a smartphone [18].

WoT is most viable as an outcome of the existing platform on which gadgets can perform. Android plays an important role at the back of WoT. A brief study of the marketplace shows that most smart gadgets function on Google's operating system, Android. If the user is familiar with Android phones, they are likewise acquainted with the working of the gadget. Presently, the sector is dealing with the Android operating system gadgets edging out iOS. In 2013, Android smartphones sold at a 4:1 ratio compared to Apple's iPhones.

The battle didn't stop with smartphones. The struggle for the pinnacle is now being fueled by means of the growing demand of interconnected gadgets. This charge is led by Android. The global WoT has been developed and controlled for the Android platform. Android are at the frontend due to various reasons:

1. Developers can create new developments for Android which work at the front end: Use of Android is increasing fast as a software program platform in general due to the fact Google chose to offer it away to builders and tool manufactures. Linux is freeware software, consequently allowing anyone to use its supply code and manipulate it for use in pretty much any device they can desire. The range of gadgets that rely upon Android to work nowadays are numerous. With these huge quantities of gadgets working on the Android OS, it is simple to see how Android works as a front end for WoT. It is simple and costs little to create gadgets for WoT, which makes them more affordable to purchasers.

2. WoT Apps drive: A device is only a device. Integration of the right app and the software program to assist it to function and perform unique responsibilities, gives it much better results. Apps are a program that makes it feasible to apply WoT gadgets. Presently Android is the biggest app platform application. In December 2016, more than 2.6 million applications were hosted by the Google Play Store. The Android platform provides WoT as the leading technology.

3. WoT is based on Java language: WoT gadgets work on the Java language. The WoT market is used by the Android platform that consequently provides an opportunity for improving software applications. Android permits Java language to be used in a way that allows improvements instead of using embedded Java-devoted gadgets.

11.4 Utility of Android in WoT

To recognize the utility of Android in WoT, it is important to recognize the WoT atmosphere and Android's role in it. Some important terms are:

1. Sensors: Sensors hit upon bodily houses together with temperature and transmit virtual signals. The majority of hardware vendors depend upon particular domain names together with Linux, Android and Windows OS. The recognition and accessibility of Android OS make it an easy winner in this region. The truth is that Android is freeware and may be used to make any tool, making it a famous preference for tool developers.

2. Transfers of data: There is a component which supports the switch of data from the sensor devices. MQTT and XMPP are the only two options available. Android helps each freeware application for simulation. Use of libraries is done through the Windows, Linux and Android operating systems.

3. Devices: A processor with a running device which is adaptable with components of WoT ecosystem. This can be a small and transportable gadget that does not use high energy, but can, however, provide uninterrupted connectivity. In most cases, cheaper Android gadgets are selected for this work. Android gadgets meet the necessities to assist various types of sensors. Many tutorials are available to assist developers to work in Android.

4. Programs: There is a need for an application which accepts the records and develops them. It should take the form of a widespread Linux Server. Information is accepted by the server, which decodes the data and makes the procedures for this. This information may be useful for next analysis.

There is not any confusion that Android OS is the primary riding pressure in back of WoT gadgets. The Android OS is necessary if a user wants to construct WoT applications.

11.5 Need for Android to Work with WoT

The "Web of Things" (WoT) are elementary regular objects which have the connectivity to the internet and device sensors which can acquire, send and accept information. Ubiquitous functionality has an effect on the Android developer, the same as smartphones which are based around an increasing number of sensors – for the entirety, like mobility, audio, temperature and contact – therefore gadgets around the house will work. A refrigerator might also have a scanner for barcodes so it knows when the user runs out of milk. These gadgets can remind the car, in a case when the user is close to a supermarket, to pick up some 2%. The Android developer must have knowledge about coding for an WoT-enabled smartphone; however, there are a few reasons to have Android software developers for developing the code for the Internet of Things:

1. Affects the job: Android packages don't do a lot without outside contact; they are required to speak to items. Gadgets must be highly linked with internet applications developed by users to speak with WoT devices. Users would possibly assume this and get away with it, due to the gadgets' communication with the server and also the user only obeys the server. Safety and overall performance are the

valuable necessities for smartphone Android software apps communicating with WoT gadgets with neighborhood networks or connected to WoT devices with one or more than one fairly vital features.

2. Skills needed to work as an Android developer: In the past, to start out with Internet of Things, sufficient information was required about protocols of hardware, networking and huge manipulation, among other matters. Instead of this information remaining treasured, tutorials and documentation exist created particularly for new beginners that are intended to take away a variety of boundaries. New developers of WoT have created groups to assist this. Now there is a tremendous array of guidance, blog posts and videos to be had on the internet to fall back upon. WoT kits are easily available which also have beginner guides and walkthroughs permitting users to develop global Morse code flashing LEDs in no time.

3. Acceptance of Android permits to mold: Understanding an era early helps to result in the "visit" man or woman in workplace; there is massive impact on the enterprise generation. Adoption may be difficult in the early stages, but more suitable within the final aspects. This acceptance is greater in terms of organizations (or the user may master it for their personal career). Learning the tough manner gives solid basis of understanding to grow from while the technology evolves and enhances users. The user has an immediate impact on the route and approach and WoT technology which is used.

4. Use of WoT in Google: As in the old adage, "a person wouldn't leap off a cliff simply because everyone is doing it," but Google using a WoT approach is accepted as right, as there may be a product space. Brillo and Weave – minimum degree technologies for interaction between gadgets and sensors – were introduced by Google. They provide a standard for WoT. Instead, most producers are developing WoT gadgets with their own protocols and stacks, which may imply gadgets are not capable of interaction. Google is allowing Weave (and Brillo as a standard) to be an upgradable protocol to permit verbal exchange and data sharing throughout developers. It provides a unique platform, better enjoyment for the purchaser and, optimistically, higher adoption. For a developer, it means an enjoyable, doubtlessly beneficial, new product space.

5. New software learning is always gaining: Knowing what's around the corner keeps the user on top of their game. If the Internet of Things does not work out user can at least have learned something. Even if the developer is on a project like Ara1, the developer must be able to observe what users have learnt about dispersed verbal exchange in WoT with Ara modules. For a developer, mastering something novel is in no way a bad thing and can most effectively increase activity.

11.5.1 Steps to Work with Android in WoT

1. At first users will want a PC to code, collect and install software to a WoT tool. For special WoT gadgets, a super-computer and operating system must be the suggested machine to apply (i.e. Windows, Mac and Linux OS). Linux machine has the most compatibility throughout WoT [18].

2. Next, the user can select their desired platform. It is the underlying hardware with a view to running sensors and gathering the information. This chapter is not comparing these items, so the following is a listing of five options:

- Board of Intel Edison
  - Brillo C++
- Board of Qualcom Dragon
  - Brillo C++
- Board of Arduino
  - C++
- Raspberry Pi
  - Scratch, Python (many other languages supported)
- BBC Micro
  - Python, JavaScript, drag & drop

After the selection of platform, the user will want to connect with a computer (that will use code for this). A few boards come with some constructed sensors devices, but the user furthermore may want a way of connecting greater sensor gadgets to the platform. There are five important things commonly required:

- Power cable
- USB cable
- Breadboard
- Breadboard wires
- Sensors, LEDs, resistors and buttons

11.6 Proposed Tool for Android Security

Our proposed tool can be used to automate data extraction from APKs that may be further used to perform analysis using machine learning algorithms. The tool is made in the Python language. AndroZoo [19] has provided a list of APKs on their website, which can be downloaded only by getting the

Application Program Interface (API) key from their owner. First, one by one, the APKs are downloaded by the tool. Each APK is then extracted using APKTool which is available for different platforms. Mainly the extracted file, namely Android Manifest.XML, is traversed and the features like permissions and intents are saved into a CSV file. For features like Restricted API Calls, the Classes.dex file is extracted using JADX Tool. After extracting the Classes.dex file, we get .java files of the Android application. From these classes, restricted API calls can be examined to see if they are present or not and they can be added to the dataset. Now the dataset formed after performing the above task can be used to apply various machine learning algorithms for malicious application detection. The methodology followed for dataset creation is explained below with the help of a diagram in Figure 11.2:

APK Downloading: The dataset of APKs from AndroZoo is a labeled data which consists of nearly 5 lacs of malicious as well as benign applications. The APKs can be accessed only with the help of an API Key, which would be provided by its owner. Once the API Key is obtained, we can download each of the APKs from their server. The description of all the APKs is given in a Descriptor.csv file which is available on their website and it contains the following columns:

1. SHA256
2. SHA1
3. MD5
4. Apk_size
5. Dex_size
6. Dex_date
7. Pkg_name, vercode
8. Vt_detection, vt_scan_date

The columns which are useful for our analysis are SHA256, which uniquely identify each of the application, and vt_detection, which provides a number, stating the total of antiviruses that have marked the application as malicious. Simply, if vt_detection value is 0, then it is a benign application; otherwise any value greater than 0 makes it a malicious application. The APKs can be downloaded by sending a request to the following URL:

https://androzoo.uni.lu/api_doc

Here ${APIKEY} is the API Key, which is provided by the owner of AndroZoo, and ${SHA256} is the specified SHA256 id obtained from the CSV file of all the APKs description. Thus, we can repeatedly go over the list and download each of the APKs with their unique SHA256 id. If the value of vt_detection is greater than 0, then we can mark the specific application as malicious

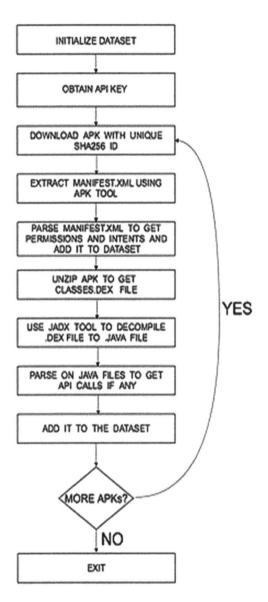

FIGURE 11.2
Process for data extraction from APK.

and if it is equal to 0 then we can mark the specific application as benign. Figure 11.3 depicts a snapshot of the actual tool where the user has to mention the following things:

1. API Key provided by AndroZoo
2. Total APKs to be taken for formation of dataset

3. Path of Descriptor.csv

4. Path where dataset has to be saved

When the user clicks on the start button, the Tool goes through the steps given in Figure 11.2. The dataset will be created when the process terminates.

An APK being downloaded is shown in Figure 11.4 which would be running in the background.

Android Manifest Extraction: After we successfully download the APK file, AndroidManifest.xml is extracted from it using a tool called APKTool [20,21]. This tool can easily extract the files stored in the APK with the commands available for it. Once we have extracted the AndroidManifest.xml we use the inbuilt Python library called xml.etree.ElementTree to parse AndroidManifest.xml. Each of the feature sets, mainly permissions and various intents, are extracted and stored using the Pandas [22] library, available in Python, in a CSV file as dataset. This dataset can actually be used for analysis afterward.

The main process of adding features like permissions and intents is done by first initializing the CSV file. Each column represents a unique feature like permissions and intents whereas each tuple represents a particular APK. The table comprises two values, that is, 0 and 1, where 0 represents a particular feature being absent in that APK, whereas 1 represents a particular APK containing that feature. If a particular feature obtained is not present in the table, a new column is created for that feature with all previous tuples valued as 0 and 1 for the present APK in consideration. This process continues for the total number of APKs considered by the user (Figure 11.5).

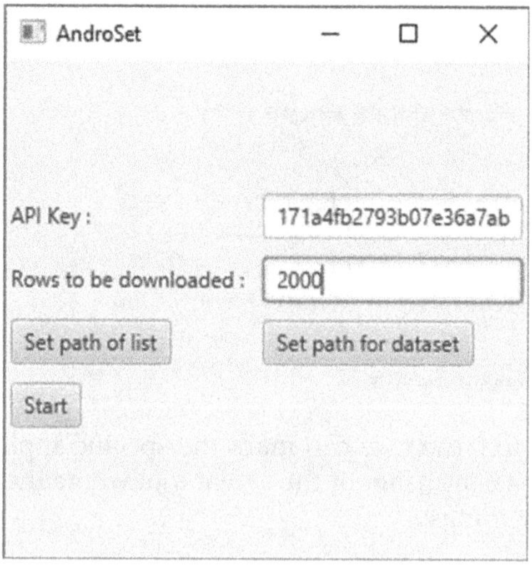

FIGURE 11.3
Snapshot of the actual tool.

FIGURE 11.4
Downloading the APK file.

Decompiling DEX code to Java: DEX files are files which contain the actual source code for the application. These are stored inside the Classes.dex file and thus we have to decompile it to get the source code. We can then perform a sweep over the various source files to determine if a particular application calls an API of Android which is suspicious or restricted. We can also determine all the network addresses which are used to communicate over the internet and determine if a particular application sends data to an unknown server.

The APK is extracted using Python's inbuilt library zipfile. Once the APK file is extracted, a small Java decompiler is run to decompile the .dex file into .java files. The decompiler is called JADX [23]. It first extracts the .class files inside the .dex file and then decompiles them into .java. Once the decompilation is over the features are extracted from the decompiled .java. The main feature set consists of API calls which are suspicious such as:

1. API calls which are used for accessing sensitive data, such as getDeviceId() and getSubscriberId()

2. API calls that can communicate over the network, for example setWifiEnabled() and execHttpRequest()

3. API calls for sending and receiving SMS messages, such as sendTextMessage()

4. API calls frequently used for obfuscation, such as Cipher. getInstance()

All these API calls along with network addresses are extracted from the dex file and converted into a feature inside the CSV being built. Now this CSV

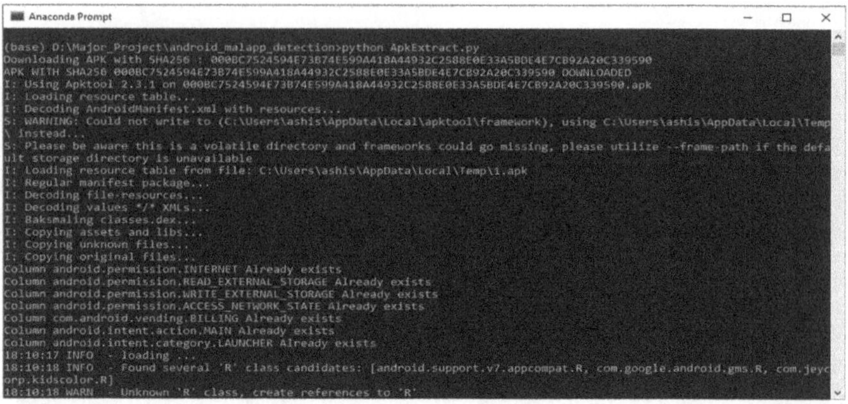

FIGURE 11.5
APK extraction using the APKTool.

FIGURE 11.6
Decompiling DEX code to Java.

file can be used as an input to perform the various machine learning algorithms, as this contains all the features extracted from the APKs (Figure 11.6).

11.7 Results

After performing the above steps we will have a dataset of the total number of applications that we had applied the above process on. The dataset would appear somewhat like Figure 11.7. Actually, the figure shown is just a

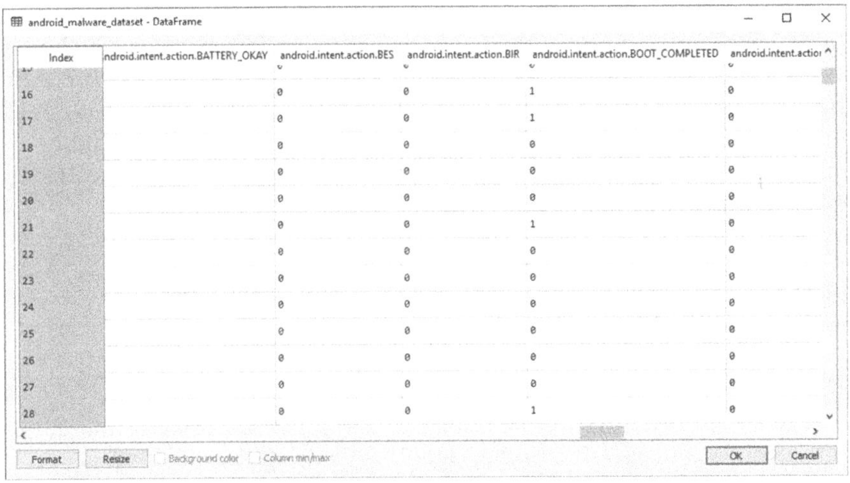

FIGURE 11.7
Snapshot of the dataset showcasing features of the applications.

snapshot of 957 rows and 1,927 columns. The database would be collection of 0s and 1s as the permission, or any feature, would either be present in a particular application or not. Each row denotes an application and each column is a particular feature which would be used during the analysis phase. It is a type of binary output as the result can be either 0 or 1. Now this dataset would vary on how many applications are considered for analysis. After getting this dataset, researchers interested in malware analysis can use various methods and strategies to predict whether an application is malicious or not.

11.8 Conclusion

Android malware is a very fast-growing threat. Classic defenses such as ant viruses have failed to cope with the amount and diversity of malware spreading in the application market. Thus, here is a very effective method to counter the malicious applications. And it makes use of data analysis techniques and machine learning. These techniques have been very efficient and useful with respect to malicious app detection. Some famous tools like DREBIN and DroidMat have been very effective and scalable for the detection process. But the scope for more productive and efficient tools is still demanding. The proposed tool thus will provide every researcher with their own custom-made dataset having the features they want to study for the applications. All they want is the APK file of an application and after that the whole work would be performed by this automated tool until the formation of a dataset in the form of CSV file. After the dataset's creation, the researcher can perform the study of analyzing in whatever way he/she wishes. The tool just requires four things to be provided at the start. After that, the CSV file would be created with the permissions in the column list and each tuple denoting an APK file. All the remaining cells would be the value, none other than 0 and 1. This dataset then could be easily used for the analysis using different approaches.

References

1. Menn, J. (2011). Smartphone shipments surpass PCs. Retrieved from http://www.ft.com/cms/s/2/d96e3bd8-33ca-11e0-b1ed-00144feabdc0.html.
2. Tam, K., Feizollah, A., Anuar, N. B., Salleh, R., & Cavallaro, L. (2017). The evolution of android malware and android analysis techniques. *ACM Computing Surveys (CSUR)*, 49(4), 76.
3. Pandita, R., Xiao, X., Yang, W., Enck, W., & Xie, T. (2013, August). WHYPER: Towards automating risk assessment of mobile applications. In *USENIX Security Symposium* (Vol. 2013).
4. Felt, A. P., Finifter, M., Chin, E., Hanna, S., & Wagner, D. (2011, October). A survey of mobile malware in the wild. In *Proceedings of the 1st ACM Workshop on Security and Privacy in Smartphones and Mobile Devices* (pp. 3–14). ACM.
5. Wei, X., Gomez, L., Neamtiu, I., & Faloutsos, M. (2012, December). Permission evolution in the android ecosystem. In *Proceedings of the 28th Annual Computer Security Applications Conference* (pp. 31–40). ACM.
6. Lindorfer, M., Neugschwandtner, M., Weichselbaum, L., Fratantonio, Y., Van Der Veen, V., & Platzer, C. (2014, September). Andrubis--1,000,000 apps later: A view on current Android malware behaviors. In *2014 Third International Workshop on Building Analysis Datasets and Gathering Experience Returns for Security (BADGERS)* (pp. 3–17). IEEE.
7. Gartner. (2017). Gartner says worldwide sales of smartphones recorded first ever decline during the fourth quarter of 2017. Retrieved from https://www.gartner.com/en/newsroom/press-releases/2018-02-22-gartner-says-worldwide-sales-of-smartphones-recorded-first-ever-decline-during-the-fourth-quarter-of-2017. Last visited 19/5/2018.
8. Li, J., Sun, L., Yan, Q., Li, Z., Srisa-an, W., & Ye, H. (2018). Significant permission identification for machine learning based Android malware detection. *IEEE Transactions on Industrial Informatics*.
9. Chandola, V., Banerjee, A., & Kumar, V. (2009). Anomaly detection: A survey. *ACM Computing Surveys (CSUR)*, 41(3), 15.
10. Bu, K., Xu, M., Liu, X., Luo, J., Zhang, S., & Weng, M. (2015). Deterministic detection of cloning attacks for anonymous RFID systems. *IEEE Transactions on Industrial Informatics*, 11(6), 1255–1266.
11. Cruz, T., Rosa, L., Proença, J., Maglaras, L., Aubigny, M., Lev, L., ... & Simões, P. (2016). A cybersecurity detection framework for supervisory control and data acquisition systems. *IEEE Transactions on Industrial Informatics*, 12(6), 2236–2246.

12. Arp, D., Spreitzenbarth, M., Hubner, M., Gascon, H., Rieck, K., & Siemens, C. E. R. T. (2014, February). DREBIN: Effective and explainable detection of Android malware in your pocket. In *Network and Distributed System Security Symposium (NDSS)* (Vol. 14, pp. 23–26).

13. Li, Z., Sun, L., Yan, Q., Srisa-an, W., & Chen, Z. (2016, October). DroidClassifier: Efficient adaptive mining of application-layer header for classifying android malware. In *International Conference on Security and Privacy in Communication Systems* (pp. 597–616). Cham: Springer.

14. Allix, K., Bissyandé, T. F., Klein, J., & Le Traon, Y. (2016, May). AndroZoo: Collecting millions of android apps for the research community. In *Proceedings of the 13th International Conference on Mining Software Repositories* (pp. 468–471). ACM.

15. Wang, S., Yan, Q., Chen, Z., Yang, B., Zhao, C., & Conti, M. (2017). TextDroid: Semantics-based detection of mobile malware using network flows. In *IEEE INFOCOM*.

16. Idika, N., & Mathur, A. P. (2007). *A Survey of Malware Detection Techniques*. Purdue University, 48.

17. Barrera, D., Kayacik, H. G., van Oorschot, P. C., & Somayaji, A. (2010, October). A methodology for empirical analysis of permission-based security models and its application to android. In *Proceedings of the 17th ACM Conference on Computer and Communications Security* (pp. 73–84). ACM.

18. APKTool. https://ibotpeaches.github.io/Apktool/. Last visited 19/5/2018.

19. DEX to Java Decompiler. https://github.com/skylot/jadx. Last visited 19/5/2018.

20. Developers, A. (2011). *What Is Android*.

21. Brahler, S. (2010). Analysis of the android architecture. *Karlsruhe Institute for Technology*, 7(8).

22. McKinney, W. (2012). *Python for Data Analysis: Data Wrangling with Pandas, NumPy, and IPython*. Newton, MA: O'Reilly Media, Inc.

23. https://examine.Sparkfun.com. https://www.sparkfun.com.

Index

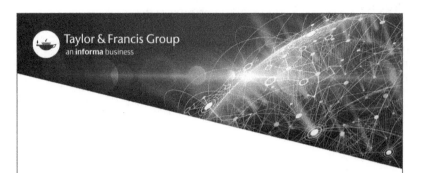